Russia's Federal Relations

The development of centre-regional relations has been at the forefront of Russian politics since the formation of the Russian state and numerous efforts have been made by the country's subsequent rulers to create a political model that would be suitable for the effective management of its vast territory and multiple nationalities. This book examines the origins, underlying foundations, and dynamics of the federal reforms conducted by President Putin throughout the eight years of his presidency. It offers a comprehensive analysis of the nature of Russia's federal relations during this period, as well as an examination of factors that led to the development of the existing model of centre-regional dialogue.

The book discusses how and why the outcomes of most domestic reforms and policies significantly vary from the initial intentions envisaged by the federal centre, and argues that – despite a range of positive developments – the reforms resulted mainly in a redistribution of powers between the two levels of government rather than in a fundamental rethinking of centre-regional relations towards genuine federalism. Chebankova argues the institutional structure of Russian federalism has become overly-centralised, monolithic, and largely authoritarian. There is a significant divergence between Russia's monocentric model of federalism and those models practised in other federal systems – which are arguably more successful in meeting democratic goals. Overall this book provides a thorough evaluation of the present state of Russian federalism and its political landscape.

Elena A. Chebankova is Research Fellow in Politics at Wolfson College, University of Cambridge, UK.

Russia's Federal Relations

Putin's reforms and management of the regions

Elena A. Chebankova

Routledge
Taylor & Francis Group

LONDON AND NEW YORK

First published 2010
by Routledge
2 Park Square, Milton Park, Abingdon, Oxon OX14 4RN

Simultaneously published in the USA and Canada
by Routledge
270 Madison Ave, New York, NY 10016

Routledge is an imprint of the Taylor & Francis Group, an informa business

© 2010 Elena Alexandrovna Chebankova

Typeset in Times by Wearset Ltd, Boldon, Tyne and Wear
Printed and bound in Great Britain by TJI Digital, Padstow, Cornwall

British Library Cataloguing in Publication Data
A catalogue record for this book is available from the British Library

Library of Congress Cataloging in Publication Data
Chebankova, Elena A.
Russia's federal relations: Putin's reforms and the management of the regions/Elena A. Chebankova.
p. cm. – (BASEES/Routledge series on Russian and East European studies; 63)
Includes bibliographical references and index.
1. Regionalism–Russia (Federation) 2. Central-local government relations–Russia (Federation) 3. Russia (Federation)–Politics and government–1991- I. Title.
JN6693.5.R43.C46 2010
320.447'049–dc22

2009026901

ISBN10: 0-415-55961-8 (hbk)
ISBN10: 0-203-86192-2 (ebk)

ISBN13: 978-0-415-55961-4 (hbk)
ISBN13: 978-0-203-86192-9 (ebk)

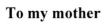

To my mother

Contents

Tables and figures

Tables

Map

Figure

Preface

By the end of the Yeltsin presidency, Russia's centre-regional relations had been in need of serious revision. It comes as no surprise that federal reform represented the first step taken by Vladimir Putin upon assuming office. The problems faced by Putin in the regional dimension were enormous, while the choice of tactics and strategies aimed to deal with these issues could have been decisive. The differing answers to Russia's territorial problems could have predefined her future existence as a single state and predetermined her successful transition to democracy.

This book examines the evolution of federal reforms during the eight years of Putin's presidency and the potential obstacles that lay on the road towards establishing an effectively functioning model of centre-regional relations. In pursuit of this task, I adopt a theoretical framework that views modern federalism in terms of the two mutually reinforcing and interlinked pillars: the structure and the process. The theoretical literature on federalism has long agreed that both these dimensions are fundamental for the stable evolution of federal states. For the working of a state's legal-institutional system strongly depends on the commitments of its composite actors to operate these structures in a manner consistent with the letter and spirit of federalism and on their adherence to the principles of reciprocity, mutuality, and interdependence.

Academic literature on Russian federalism, however, customarily concentrates on the institutional aspects of federal governance and on the influence of the country's territorial problems on her democratic transition. This often takes place at the expense of discussing the ethical and socio-political dimensions of federal relations, as well as their real functional nature. This book intends to make a difference in this field by drawing a careful distinction between the structure and the process of Russian federalism and by discussing the emerging gulf between these two elements as an important political problem that could seriously influence the developmental dynamics of the extant federal state.

My analysis of the reforms' results demonstrates that, in the institutional sphere, the centre managed to halt centrifugal tendencies and re-establish the authority which had been eroded during the Yeltsin era. At the same time, I argue that, despite a range of positive developments, the reforms resulted mainly in a redistribution of powers between the two levels of government and not in a fundamental rethinking of centre-regional relations towards genuine federalism. The centre placed itself at the top of the new 'regional hierarchy' and enhanced its control over the socio-economic activity of the regions, effectively turning them into political 'clients'. The institutional structure of Russian federalism has therefore become overly centralised and largely authoritarian.

Another strand of my discussion, however, concentrates on the real process of Russia's federal relations. I argue that, with the erection of the rigid framework, the process of Russian federalism, the real federal bargaining, has not disappeared from the country's political landscape. The centralisation of the federal institutional structure prompted the regions to embark on covert political activity aimed at relaxing the existing institutional system and at effecting devolutionist structural moves. This regional drive towards decentralisation, however, assumed a variety of disguised forms, in which the real fabric of federal bargaining was hidden behind a monolithic institutional façade.

As a result of these inconsistencies, Russia has made no significant steps towards achieving greater federal integration based on functional co-operation between national and regional institutional structures. Neither has it established a system of incentives which could encourage central and regional elites to maintain their dialogue regardless of external pressures and changes in ruling political circles.

Successful resolution of such federal problems is of paramount importance for the development and modernisation of the Russian state. Given Russia's vast territory and the multiplicity of her indigenous nationalities, any significant domestic policy issue invariably raises the question of how such initiatives should be implemented in the regions. This book will show that the outcomes of the country's domestic reforms and policies significantly vary from the initial intentions envisaged by the federal government. For the regional reality transforms and adapts these dynamics to unique local conditions that are often influenced by reactionary cultural and historic constraints. This holds true for a wide spectrum of initiatives, including those pertaining to the broad aims of democratisation, effective economic governance, and socio-economic modernisation.

This work was initially intended as a doctoral dissertation, which was completed at King's College, Cambridge in May 2004. I later expanded this project to include a discussion of the second term of Putin's presidency. This came in the form of various research articles that compose a large part of this book. I am indebted to many people who at various points and in various capacities influenced my ideas and encouraged me to persevere with this project.

In pursuing the research, I have been very fortunate to have been supervised by Dr John Barber. His professional advice and the academic attention he has given to my work could hardly be overestimated. More importantly, he often went far beyond his academic duties and provided me with much needed moral support and guidance. My thesis examiners Professor Archie Brown and Professor Richard Sakwa have long been a source of academic inspiration and set scholarly standards to which I will always aspire. I would also like to thank Dr David Lane who, at later stages, provided me with more than generous professional guidance, help, and support. I am also very grateful to Professor Stephen White who has given much support and encouragement during my post-doctoral research. My thanks also go to Dr Harald Wydra whose approach to the study and teaching of Russian politics has been particularly stimulating.

Among the many colleagues and friends that I met during my fieldwork in Russia, I owe a special debt of gratitude to Alla Yevgenyevna Chirikova. My research could not have been half as productive without her help. She was an invaluable source of contacts, wisdom, information, and knowledge on Russia's regions. It is through my conversations with Alla Yevgenyevna that my vision of Russian regional politics broadened and my principal arguments began to take shape. Olga Borodinova in Yaroslavl assisted with scheduling and organising interviews and put me in touch with many people who largely contributed to my understanding of the subject.

I have to particularly acknowledge my intellectual debt to Yakov Pappe, who has been more than generous with his time, and provided me with an insight into regional economic development. I shall always aspire to the high standards of his work and the rigour of his political analysis. Also, interviews with Alexei Zudin, Natalya Zubarevich, and Natalya Lapina have helped significantly in shaping my vision.

I am very thankful to my friends who supported me throughout this period, everyone in their unique capacity: Jonathan Knightley, Francois Mann Quirici, Sébastien F. Brack, Andrew Evans, Floris Van Nierop, Paulius Kuncinas, Paul Goode, William A. Clark, Maude Vanhaelen, Richard Gale, Katerina Kocourek, Astrid Strohbach, Thomas D'Andrea, Sabine Chaoche, Roberto Polito, Marina Baranovsky, Maurissio Carradini, Grace-Ann Duncan, Vitaly Iliashenko. I am very grateful to my friend Bryan Cain who proofread some parts of this work. I also must thank Victor Burlakov, Asel Sartbaeva, and Julia Hudson for our long conversations on politics, economics, and theology at Linacre College, Oxford.

Friends and people I met in Russia have been a great source of support and inspiration. Pavel Konstantinov, Robert Agababyan, Natasha Bibikova (Gorshkova), Evgenii Burlov, Ilya Dudkov, Mikhail Kartashev, Oleg Kovalev, Natalya Kopteva, Olga Kulikova, Tatyana Lavrushechkina, Alexandr Lartsev, Dmitry Mamatov, Andrei Soldatov, Evgenii Sorokin, Olga Tanenberg, Yurii Tikhomirov – all made my visits to Russia very special and wonderful.

My special thanks go to the family of James Robertson, Jeff Robertson, and Janet Robertson, who were always there for me and without whom my academic work would have been particularly tough. My step-aunt Raisa Sorokina has always been a great source of inspiration and optimism. I often thought of her wisdom while working on my manuscript in Oxford, and always looked forward to seeing her while in Russia. My aunt Lyuba and uncle Sasha must also be mentioned as giving me unconditional support and always made me feel so much at home in Russia. I greatly enjoyed my conversations on politics with my cousin Igor Mistyukov. As a member of the Moscow Regional Board of Lawyers, he very kindly introduced me to some regional politicians and helped me in forming my view on regional developments.

I am most grateful to my parents: my mother Vera, my stepfather Anatoly, and my father Alexander. My stepfather has always understood the complexities of my life in research and was keen to assist me in his professional and personal capacities. Many interviews in the Moscow Regional Duma have been scheduled with his kind assistance. Long conversations, and often-heated debates with all my family members, provided me a very 'Russian' view on the subject. Without this input, I would have not been able to raise the many important questions discussed in this work. Finally, nothing is possible without my mother, who endured the hardships of my academic pursuits and has always been my greatest source of support. This work is dedicated to her, with everlasting love and gratitude.

Elena Chebankova
Cambridge, 2009

Acknowledgements

The Author and Publishers would like to thank Taylor & Francis Journals for kind permission to reprint in *Russia's Federal Relations,* extracts from the following articles previously published in their journals <www.tandf.co.uk/journals>:

'The Limitations of Central Authority in the Regions and the Implications for the Evolution of Russia's Federal System', *Europe-Asia Studies*, Vol. 57, No. 7, November 2005, pp. 933–49.

'The Unintended Consequences of Gubernatorial Appointments in Russia, 2005–6', *The Journal of Communist Studies and Transition Politics*, Vol. 22, No. 4, December 2006, pp. 457–84.

'Putin's Struggle for Federalism: Structures, Operations, and the Commitment Problem', *Europe-Asia Studies*, Vol. 59, No. 2, March 2007, pp. 279–302.

'Implications of Putin's Regional and Demographic Policies on the Evolution of Inter-Ethnic Relations in Russia', *Perspectives on European Politics and Society*, Vol. 8, No. 4, December 2007, pp. 439–59.

'Adaptive Federalism and Federation in Putin's Russia', *Europe-Asia Studies*, Vol. 60, No. 6, August 2008, pp. 995–1015.

Theoretical parts of these articles have been used in Chapter 1 of this book; empirical materials of these articles have been included in Chapters 4, 6 and 7.

Abbreviations and glossary

AGCO	Anglo-American Agricultural Company
AO	Autonomous District (*Avtonomnaya Oblast*)
BP	British Petroleum
CFD	Central Federal District
CIS	Commonwealth of Independent States
CPRF	Communist Party of the Russian Federation
CPSU	Communist Party of the Soviet Union
CSR	Corporate Social Responsibility
ER	*Edinaya Rossiya* (pro-presidential political party)
FOM	Fund for the Study of Public Opinion
FSB	Federal Security Service (successor to KGB)
FZ	Federal Law (*Federalnyi Zakon*)
GAZ	Volga Automobile Factory
Gazprom	Russia's state gas monopoly
Gazpromneft	Russia's oil company
GDP	Gross Domestic Product
Golos Rossii	Voice of Russia (electoral bloc)
gosudarstvo	state
Grazhdanskaya Sila	Pro-Kremlin political party of the liberal right wing spectrum led by Mikhail Barshchevskii
IBG	Integrated Business Group
IMF	International Monetary Fund
Interros	Integrated business group headed by Vladimir Potanin
Irkutskenergo	Electricity Company in Irkutsk Oblast
KGB	State Security Committee (Soviet Intelligence Service)
krai	large region
Kraz	Krasnoyarsk Aluminum Factory
LDPR	Liberal Democratic Party of Russia
Lukoil	Oil company led by Vagit Alekperov
MDM	Integrated business group led by Andrei Melnichenko
MEDEF	French Movement of Enterprises
Mezhprombank	Bank headed by senator Sergei Pugachev
MOD	Ministry of Defence

MP	Member of Parliament
MVD	Ministry of Interior
NATO	North Atlantic Treaty Organisation
NGO	Non-governmental organisation
NLMK	Novo-Lipetskii Metallurgy Plant headed by Vladimir Lisin
nomenklatura	Communist system of appointment; representatives of high official positions within Communist hierarchy
Nornikel	Norilsk Nickel (mineral mining company)
NPSR	National Patriotic Union of Russia (political movement)
obkom	Regional Committee of the Communist Party of the Soviet Union
oblast	region
okrug	large territory or district, region
OMZ	United Machines Plant headed by Kakha Benukidze
OPORA	Union of Russia's Entrepreneurial Organisations
oprichnina	Ivan IV's policy of ruling part of Muscovy through terror
Otechestvo	Fatherland (electoral bloc)
OVR	Fatherland-All-Russia (electoral block)
polpred/polpredy (pl.)	plenipotentiary representative of the president of the Russian Federation in the Federal District
propiska	the system of registration of residence
raion	district
RAN	Russian Academy of Science
RAO RZhD	Russian Railways (state transportation monopoly)
RAO UES	United Electricity Systems (state electricity monopoly)
RF	Russian Federation
RLK	Russian Leasing Company
Rodina	Motherland (electoral block)
ROMIR	Independent Research Centre for the Study of Public Opinion
Rosneft	Russia's state-owned oil company
RSFSR	Russian Socialist Federative Soviet Republic
RSPP	Russian Union of Industrialists and Entrepreneurs
Rusal	Russian Aluminium headed by Oleg Deripaska
Sberbank	Savings Bank of the Russian Federation
Sibneft	Oil company headed by Roman Abramovich
Sibur	Oil refining company formerly headed by Iakov Golodovskii
siloviki	representatives of legal enforcement ministries
SPS	Union of Right Forces (political party)
SR	Spravedlivaya Rossiya (left-wing party of power)
STS	State Tax Service
Surgutneftegaz	Oil company headed by Vladimir Bogdanov

Svyazinvest	Telecommunications monopoly
TNK	Tyumenskaya Neftyanaya Companiya (oil company)
Transneft	Russia's oil transporting monopoly
UGMK	Ural Mining and Metallurgical Company
Vsya Rossiya	All Russia (electoral bloc)
VTsIOM	All-Russian Centre for the Study of Public Opinion
Yabloko	Political party led by Grigorii Yavlinskii
Yukos	Russian oil company formerly headed by Mikhail Khodorkovskii
Zemstvo	territorial unit of late Imperial Russia

Note on spelling

As it is virtually impossible to transliterate all Russian names consistently, except by means of a variety of specialised annotations which would necessitate their own glossary, the Modified Library of Congress system has been adopted to ensure consistency. The letters 'ia' and 'iu' have been transliterated as 'ya' and 'yu'. In addition, some Russian names have been anglicised.

1 Methodology, theoretical considerations, and the structure of the study

Russia's post-Soviet transformation has been marked by a significant difference between implemented institutions and political practices. Many progressive liberalising initiatives that were borrowed from the West faced serious difficulties at the execution stages, while the final outcomes diverged markedly from the original intentions. The same can be said of the efforts to centralise and constrict political activity. The resulting reality has often proved to be more diverse, complex, and much less manageable than initially anticipated by the policy architects.

Putin's federal reform of 2000–8 provides a plethora of opportunities for studying the evolution of institutional frameworks and their relations with societal and cultural realities. The programme created seven federal districts as a new administrative structure, instituted a new position of plenipotentiary representatives (*polpredy*) to supervise these formations, modified the method of constituting the Federation Council, established the State Council as a forum for dialogue between the president and the regional governors, and introduced a range of new measures for enforcing compliance with federal law in the regions.

This package was later complemented by a series of laws further reinforcing the position of the centre vis-à-vis the regions. These concerned the introduction of a new inter-budgetary system, the restructuring of the Ministry of the Interior, the establishment of a Legislative Council, and modification of the administrative processes related to regional and federal elections and to the activities of political parties in the regions. Business-state relations in the regions have come under restructuring with the introduction of new legislation aimed at the regulation of taxation, licensing, and the functioning of regional economic institutions. The processes of inter-regional integration have been accelerated with the revision of the borders of some federation subjects and the introduction of additional economic and political stimuli. Official accounts of the intention behind this re-centralisation campaign included strengthening of the principles of federalism embodied in the Constitution, the harmonisation of federal and regional legislation, and the elimination of a number of legal contradictions relating to the functioning of the Federation Council.[1]

Just like with other policy areas, these initiatives represented an amalgam between progressive democratic and authoritarian-arbitrary undertakings. In both cases, the end results varied from the initial plans. Indeed, restrictive measures that were intended to enable the Kremlin to secure a tighter grip over the regional politics did not lead to a complete unification of Russia's regional space. Rather, they altered the centre-regional balance of power in the Kremlin's favour only in certain areas with certain reservations contingent on a range of external conditions. A similar logic applies to a spectrum of initiatives aimed at modernisation of federal institutions, regional economic development, and checking the mode of business-political dialogue within the provinces. These policies have been patterned on the existing, well functioning Western models and should have brought some positive outcomes in the short-term perspective. The results of such initiatives, however, have proved to be contrary to what was initially expected and in some cases, an acceleration of pre-existing negative tendencies was observed.

This book dissects Russia's emergent federal structures and the actual political processes taking place within them. The surfacing of an apparent rift between these two categories will constitute the primary aim of this account. The existence of a regional political context is perhaps the most powerful explanation to the existence of such a split. Indeed, Russia's regional politicians almost always adapted the Kremlin's initiatives to their unique cultural, territorial, and ethnic conditions. This has led to the modification and diversion of the original intentions. Thus, no serious central policy initiative can be viewed without accounting for the regional socio-economic, socio-political, and socio-historic conditions. In those cases where constrictive measures prevailed, these local conditions were determined by the regional longing for a greater political self-determination, ethno-national pressures towards a wider recognition, as well as the sheer impossibility of sucessfully managing Russia's territorial complexity unilaterally from a single arbitrary centre. In cases of modernisation and democratic shifts, path-dependence and cultural factors played their role. Thus, many progressive initiatives have been introduced arbitrarily, partly undermining their intended liberalising purpose.

This chapter is theoretical. It presents definitions and concepts through which these dynamics can be understood. I will begin this analysis with its territorial aspects and examine various facets of federal structures and processes. The discussion will examine some major tenets of federal ideology and culture, as well as the most fundamental structural pillars supporting federal institutions. I will put forward an argument that the process of real federal relations in Russia remains very intricate due to the complexity of the country's federal society and her extreme territorial diversity. At the same time, I will claim that a complex interaction has developed with the growing rigidity of federal institutional networks. It has adapted to the current structural realities and taken on some disguised non-transparent forms. This factor has resulted in the growing rift between the erected institutional structures and the functioning processes taking place within them.

The existence of such a rift invariably raises questions of a structrual change and potential liberalisation. This brings us to another important theme of this book – *political style*. *Political style* constitutes a substantial dimension of political culture and has a profound influence on *how* politicians understand and implement a broad range of initiatives. I will argue that the existence of stylistic constraints may stifle many liberalising moves by federal society, as well as governmental intentions to erect those federal institutional structures that could comply with democratic standards pursued by the world's most successful federations. In the longer term, the emergence of the gap between the structure and the process of Russia's federalism can prove to be a serious political problem. For, regardless of stylistic, cultural, or institutional constraints, the existing model of regional relations will head towards some acceptable forms of balance. This signals an unwelcome message for the sustainability of the current federal structures, as well as to the ways in which political actors operate them.

Federalism and federation

Modern federalism is universally viewed in terms of the two mutually reinforcing and inextricably linked tiers: the structure and the process (Burgess and Gagnon 1993; Friedrich 1968; Elazar 1987). King (1982, p. 6) was among the first political scientists who made the existence of these two pillars systematic. He proposes an important methodological distinction between *federalism* and *federation*, in which the former represents an ideology, a conceptual value, an organisational principle, and the latter a technical institutional arrangement (see also Watts 1999, p. 6).

The development of Russian federal relations during the post-Soviet period has been particularly marked by the complex interplay between the concepts of federalism and federation, or the structure and the process of centre-regional relations. A number of scholars have attempted to elaborate on the dynamics of their interrelationship. Ross (2002, p. 7), supported by Hahn (2004, p. 123), refers to Russia as 'a federation without federalism'. This statement, interesting from both theoretical and empirical points of view, invites further investigation. In order to accomplish this task, we first have to establish the nature and origins of these notions.

Federalism as an idea and a process

Federalism, as an idea, is based on the two most important conceptual pillars. First, it conveys covenantal, contractual relations (the word *federal* is derived from the Latin *foedus*, which means contract, compact or bargain) among the composite units of a state or society. Second, it strives to reconcile the aspects of unity and diversity, freedom and authority, creation of a wide social entity and the preservation of uniqueness of its constituent parts.

The covenantal dimensions of federalism find their origins in the covenant theory of the Hebrew Bible (Elazar 1987, p. 117). The essence of the original federal thought was that humans could become God's free and equal partners committed, at the same time, to a relationship of mutual responsibility. These ideas resurfaced and consolidated with the emergence of Reformation era theology. Leading theologians of that period, Heinrich Bullinger, John Calvin, Martin Luther, Ulrich Zwingli, stressed the idea of a covenant between God and his people. This implied that Man concludes a compact with God in order to achieve eternal love and blessing and promises in return to keep faith and uphold God's law on earth.

The seeds of theological federalism had enormous political implications, in that they raised the questions of the limitations of power in both church and state and of governmental checks and balances. A large body of Reformation-era radical political thought – Philippe Mornay, John Knox, Christopher Goodman – relied on the covenantal principles in explicating the right of individuals to resist tyrannical rulers. The logic was that, since each citizen has promised to God to uphold his law, each of these citizens has thus promised to 'root out evil and to repudiate all forms of idolatry and tyranny'. It was therefore a duty of such citizens to resist the tyrannical powers, and failure to do so would lead to a divine condemnation and be 'tantamount to breaking the league of covenant which they have sworn to God himself' (Skinner 1978, pp. 237–8; see also pp. 325–6).

This thought inspired Johannes Althusius to further articulate and secularise the existent federal idea. As Baker and McCoy (1991, p. 55) note, 'what Althusius undertook to do is to interpret all political life in terms of *pactum*, the bond of contractual union, or covenant'. Althusius held an ultimately integrative federalist view, in which humans come together in a covenantal accord to fulfil their various material and social needs and to enhance their life together. He asserts that 'humans are symbiotic beings by their covenantal nature, not separate individuals. There is a symbiosis between God and humanity, between humanity and creation, and among humans in community' (Baker & McCoy 1991, p. 52).

The most important part of Althusius's thought, however, was a tentative introduction of the territorial principles of such a covenant. Althusius insisted that covenantal relationships exist not only between Man and a centralised State or among single individuals, but also within families, which then unite into villages and towns to form provinces and proceed to commonwealths with the aim of establishing covenantal orders at all such levels (Freidrich 1968, p. 12). Therefore, speaking of modern federal systems, Elazar (1987, p. 185) observes that the 'contractual sharing of public responsibilities by all governments in the system appears to be a central characteristic of federalism'. And Freidrich (1968, p. 6) continues that 'a federal order typically preserves the institutional and behavioural features of *foedus*, a compact between equals to act jointly on specific issues of general policy'.

The second most important aspect of federal thought concerns the problem of preserving individual integrities within the given covenantal unity. In theological terms, the relations between God and humans could not be one of equals. This problem was, however, resolved through the assumption that God has 'graciously limited' Himself and, by restricting His otherwise omnipotent powers, He has granted a significant degree of freedom and integrity to humans (Elazar 1997, pp. 117 and 115).

In this context, Rufus Davis (1978, p. 3) writes that 'the idea of covenant betokens not merely a solemn pledge between two or more people to keep faith with each other, to honour an agreement; it involves the idea of co-operation, reciprocity, mutuality, and it implies the recognition of entities – whether it be persons, a people, or a divine being. Without this recognition there can be no covenant, for there can be no reciprocity between an entity and non-entity'.

Indeed, Elazar (1997, p. 33) clearly explains that 'federalism has to do with the need of people and polities to unite for common purpose yet remain separate to preserve their respective integrities'. He (1997, p. 67) further stresses the need for the covenanting federal parties to strive for a consensus in resolving their existential issues and, failing that, seek 'an accommodation that protects the fundamental integrity of all the partners'. Thus, at the heart of federalism lies a drive towards settling an eternal conflict between 'unity and diversity', which is combined, in Duchacek's (1970, p. 192) view, 'with a keen awareness of mutual dependence'.

In structural terms, Proudhon thought that federalism represents a perfect institutional reconciliation of 'liberty and authority' (King 1982, p. 40). In this sense, federalism, as Burgess (1986, p. 13) insists, is 'simultaneously sustained by the twin driving-forces of differentiation (or diversity) and integration'. The nature of federalism, therefore, can be consistent with both centralising and decentralising dynamics. King (1982, pp. 75–6) brands these dynamics as 'normative orientations' and argues that these 'can change over time within each federation and vary between them all'.

We can clearly see that federalism, as an idea and a concept, emerged from the religious thought of the Reformation era, and was further developed and secularised to give rise to important political doctrines and structures. In its original Western sense, this thought strove to accommodate the diversity of varying political entities within a single whole on a contractual covenantal basis. It comes as no surprise that the first political practice of federal relations took shape in those countries of Europe where the Reformed communities were particularly prominent – the United Provinces of Netherlands and Switzerland (Friedrich 1968; Baker and McCoy 1991). Baker and McCoy (1991) argue that the 'Reformed tradition then headed for England and Scotland, and finally settled in North America. By then the Puritans had brought with them their distinct religious, ethical, and moral traditions, which served as an organising principle for the colonies of New England'.

Federation as a structure and institution

Madison made the first formal attempt to fully transform the federal thought into the modern institutional principle and technique, and a large number of countries have followed the United States of America in organising their governments on a federal basis. This allowed devising some important structural criteria that distinguish such arrangements from other institutional systems. In this light, all federal polities are fundamentally based on the existence of a written constitution and a formal division of powers between central and regional governments. The constitution is required to ensure that the federal nature of the political system is protected by supreme jurisdictional authority and is almost impossible to change (Dicey 1967, p. 141; Wheare 1963, p. 54). This requirement is also consistent with the aforementioned ideological proposition that all federal relations must be ensured by some form of contractual, covenantal agreement. The division of powers guarantees that each level has 'some activity on which it makes final decisions' (Riker 1975, p. 101). This is proposed to safeguard the idea that all federal relations are driven by the principles of mutuality, reciprocity, and interdependence, and that the two orders of government should communicate on the basis of partnership, and not hierarchy (Elazar 1997, p. 239; Lijphart 1999, pp. 186–7). These theoretical provisions formed the basis for a wide academic enquiry. Riker (1964, p. 11), King (1982, p. 77), Dahl (1986, p. 114) and Lijphart (1984, p. 170) referred to such notions as the main departure point in exploring the nature and dynamic of federal polities.

Before embarking on a discussion of all structural components of a federation it is important to separate this concept from other multi-unit territorial states that could be mistaken for a federal polity. Therefore, we need to ascertain precise and mutually exclusive definitions concerning the institutional nature of multi-unit territorial formations. It is widely perceived that all such formations generally fall into three broad categories: federation, confederation, and devolution of power. Given that the first group is the matter of interest in this account, it is feasible to examine precise definitions of a federation while evaluating the other entities in comparative terms. And in this light, the two aforementioned principles, i.e. the method of dividing powers between regional and central governments and the institutional role of the federal constitution, have long been accepted as the most fundamental criteria distinguishing federations from other forms of institutional arrangement.

In his seminal 1946 work on federalism (republished in 1963), Sir Kenneth Wheare emphasised the importance of both these requirements. His definition of the *federal principle* (1963, p. 10) outlined the method of division of powers in which 'the general and regional governments are each, within a sphere, co-ordinate and independent' (1963, p. 10). Moreover, both governments must operate directly upon people. This requirement is conditional upon various matters such as the precise definition of the sphere of influence for both governments and problems concerning the distribution of residual powers, i.e. those not officially outlined by or absent from the constitution. Despite all these issues, the underlying logic behind this definition is very solid, in particular when the criterion proceeds to determine the rules for confederations and unitary states with devolution of power.

It is within this framework that *confederation* represents the method of dividing powers so that the general government is dependent upon the regional governments and the general government does not operate directly upon people. *Devolution of power*, on the other hand, stands for the method of governing in which regional governments are subordinate to the general government and both governments operate directly upon people (Wheare 1963). These definitions already provide the researcher with a yardstick for evaluating to what extent one or another state constructed federal institutions.

Furthermore, the role and place of the constitution in the multi-unit state institutions represents another fundamental pillar separating a federal state from confederations and unitary states with devolved power. Wheare (1963, p. 54) states that 'if a government is to be federal, its constitution... must be supreme. The terms of the agreement which establishes the general and regional governments and which distributes powers between them must be supreme and binding upon general and regional government'. While unitary states with devolution of power comply with this requirement, the constitution does not have a supreme and binding effect in confederations. Bearing this structural picture in mind, King (1982, p. 77) succinctly defines federation as 'an institutional arrangement, taking the form of a sovereign state, and distinguished from other such states solely by the fact that its central government incorporates regional units into its decision-making procedure on some constitutionally entrenched basis'.

At the same time, following the introduction of these two fundamental factors, academic focus has gradually shifted towards the nature of federal processes, and in particular the establishment of additional institutional networks that could support, enhance, and facilitate the functioning of a skeletal system defined merely by a written constitution and formally recognised division of powers. Friedrich (1968, p. 176), for example, insisted that 'federalism should not be seen as a static pattern or design characterised by a particular and precisely fixed division of powers between governmental levels. Federalism is also, and perhaps primarily, the process of federalising a political community, that is to say, the process by which a number of separate political communities enter into arrangements for working out solutions'.

A growing body of academic thought that responded to these ideas has begun to elaborate on the extent to which additional institutional networks contribute to the stable and effective functioning of federal systems. Davies (1978, pp. 143–6) claimed that it is impossible to achieve a perfect division of powers within a federation without formation of various associations that would inevitably lead to 'intended and unintended relations'. Sawer (1976, pp. 113–14), in searching for the roots of federal stability, discovers the importance of an institutional 'machinery', primarily of financial, fiscal, and political character, that is called to enhance the relations between the central and regional levels of authority. Chapman (1993, pp. 71–4) concludes that a range of supporting institutional structures, 'at times extra-constitutional and, in many cases, extra-parliamentary' must be established to ensure that the federal model is sustainable and operationally effective.

Lijphart succinctly brands these structures as 'guarantors of federalism'. He (1999, pp. 187–8; see also Duchacek 1986, pp. 188–275) primarily views a bicameral legislature with a strong federal chamber and an effectively functioning legal system as the main composite units of these supporting institutions. There is also an array of literature that elaborates on the nature of many additional structures. At least three pillars, which contribute to operationally effective federal polities, can be distinguished: (1) developed institutions of centre-regional functional co-operation; (2) a transparent system of centre-regional migration of cadres; and (3) an integrated national party system.

Many scholars (Elazar 1987, p. 184; Smiley 1972; Watts 1989) have claimed that comprehensive networks of intergovernmental co-operation foster closer political relations between the centre and the regions, make territorial democracy more 'operationally effective', and motivate regional politicians to sustain the functionality of the existing federal structure. Academic analyses of stable federations usually demonstrate that the institutional strength of these systems is sustained by the fact that regional elites' gains in maintaining stable federal order are sufficiently greater than benefits from defecting and pursuing regional sovereignty. Niou and Ordeshook (1998, pp. 279–80), in particular, note that financial-political elites often attempt to create institutions that 'render the federation's maintenance in the self-interests of each succeeding generation of political elites of the federations' subunits'. Indeed, in all consolidated democracies, politico-financial elites are motivated to maintain the existing regime because they are integrated into the context of formal democratic institutions – parties, parliaments, civil societies, trade unions – and exercise their informal and formal lobbying through these institutions. In federal democracies, political elites receive additional means of achieving their interests through intergovernmental institutional co-operation between the centre and the regions. The effective functioning of these institutions creates further incentives for the elites to maintain the existing federal order.

Indeed, most federal countries began to implement such a system during the second half of the twentieth century (Bakvis and Chandler 1987; Simeon 1985) and a large number of post-war federations enshrined these dynamics in their constitutions (Birch 1957, pp. 291–304). Switzerland, as one of the most stable and successful world federations, deserves a special mention. All levels of the country's power – federal, cantonal, and municipal – exercise close co-operation through meetings and conferences organised on an institutional and ad-hoc basis. In particular, compulsory annual meetings between central and cantonal governments, at which three representatives of the centre and three selected representatives of the regions discuss national issues, take place twice a year (Dyshekova 2003). Moreover, the new Swiss constitution of 1 January 2000 established the Council of Presidents of Cantonal Governments with the view of enabling cantonal executives to participate in the federal decision-making process, enhance co-operation between cantonal governments, and enable the cantons to use new ways and tools for cantonal and inter-cantonal partnership co-operation (Fleiner 2002, p. 118). Spain, another example, enhanced levels of federal integrity through the establishment of venues of functional centre-regional co-operation in selective economic and political matters (Börzel 2000, pp. 21–2).

The evident continuity in the careers of regional and central politicians, in accordance to Ordeshook (1996) and Filipov *et al.* (2004), ensures that sub-national executives would employ their current positions as a springboard to federal politics. In this case, they would have to take a greater interest in the national political agenda and work in co-operation with the centre, maintaining the stability of the existing federal bargain and the supremacy of federal law. Indeed, an effective process for the national promotion of regional elites is practiced by a number of successful federations such as the United States, where many presidents served as State governors during their political careers and presidential cabinet members come from important administrative and political positions in different states (Ordeshook and Shvetsova 1997, p. 34). In addition, the president officially represents a national party and his/her candidacy is proposed by a college of electors normally comprising representatives of states' party elites (Lowi and Ginsberg 2000, pp. 150–2). Furthermore, because each presidential candidate must build a winning coalition state by state, the president comes to office having strong political connections with the regions, and his incentives to co-operate with states' political elites and to promote some of them to national positions are clear to everyone in the centre. At the same time, the very possibility of employing regional politics as a springboard to national government prompts regional politicians to refrain from antagonising the centre and to resolve difficulties by relying on centripetal co-operative means regardless of the centre's ability to provide economic subsidies.

The system of integrated national parties represents perhaps the most fundamental element of centre-regional co-operation and the cornerstone of effective federal relations (Riker 1964; Kramer 1994; Filipov *et al.* 2004). This model, defined by electoral, ideological, financial, and organisational cohesiveness between national and regional party structures, is designed to bridge gaps in federal and regional policy ends. In Russia's case, the creation of a national party system is a particularly salient task. Stoner-Weiss (2002, p. 145) insists that the weakness of parties as integrating institutions means that there will be continued instability in centre-periphery relations, and therefore ineffective governance of Russia... in the absence of well-institutionalised parties (and other institutions that might effectively link centre and periphery), Russia is more likely to develop what William Riker termed a 'peripheralised federalism' rather than centralised federalism.

Indeed, political dangers associated with building an overly 'peripheralised' regional party system, were discussed in detail by a number of analysts even with the example of the already established federations. In his comparative analysis of American, German and Canadian federal systems, Ordeshook (1996, pp. 195–217) demonstrates that the existence of regional parties disconnected from national politics undermines efforts at 'gluing' a federation politically into one united system, and such non-integrated systems finally become susceptible to the emergence of secessionist trends. Ordeshook (1996) argues that separatist tendencies in Canadian provinces mainly stem from the fact that the Canadian party system is rigidly divided into regional and national units. In this system, each region has a number of unique parties, native to that particular region only, and regional parliaments that solely comprise representatives of these regional parties. Politicians representing regional and national parties do not mix and affiliation to either regional or national party is often a mutually exclusive choice that a politician has to make at the beginning of his career. The lack of connection between regional and national politics is also manifest in the fact that no Prime Minister of Canada has ever served as head of provincial government. This contrasts with the integrated system of the United States, where nearly half of all presidents held the office of state governor. In addition, the Canadian upper chamber is composed of appointed representatives from regional parties. Due to the multiplicity of these parties' agendas and the resulting inability of the senators to form strong factions such a system marginalises the political weight of the senate. This party structure often leads to the situation in which regional politicians have no national concerns, responsibilities or ambitions. Moreover, psychological links between national and regional politics do not ensure integration – it is often the case that provincial governments and politicians regard the Canadian national government as an 'alien' force (Ordeshook 1996, p. 208).

Federalism and federation in Russia

With this structural picture in mind, it is important to mention that the establishment of these institutions, both fundamental and auxiliary, is still insufficient for the achievement of a successfully functioning federation. As Livingston (1956, pp. 1–4) notes, 'a society may possess institutions that are 'federal' in appearance but may operate them as though they were something else'. Franck (1968, pp. 167–83), summarising the experiences of failed federations in West Indies, Central and Eastern Africa and Malaysia insists that 'the mere creation of federal infrastructure... cannot long contain basic political and social conflict'. He claims that in developing nations politicians must exhibit a serious level of commitment to the primary concept or the value of federation itself. Many scholars, (Ostrom 1991; Beer 1993; Burgess and Gagnon 1993; Bednar 2004; Bednar 2005) have supported these ideas by subsequently arguing that federal structures, despite the paramount importance of their mere existence, must also function in a manner that exhibits their commitment to the spirit and letter of federalism.

Thus we are now facing an important problem of interaction between federal institutional structures and federal culture, process, and ideology. This discussion allows an important proposition, namely that federalism represents not only an ideology, as we have discussed above, but also a state of society, almost its underlying nature, a set of problems that needs to be answered in a particular way. In this sense it is, and always will be, primary to a federation, which represents some form of institutional arrangement called to reflect these realities.

Livingston (1956) was a particularly avid advocate of the social origins of federations. He claims that the federation, as a form of government, emerges in response to the underlying shape of society. Watts (1966, p. 16) continues that federal states are produced by social infrastructures. Burgess (1986, p. 13) further asserts that federations 'are not magicked into existence ... they were created to solve particular problems at a particular point in each state's development'. In this light, King's (1982, pp. 74–6) conclusions that 'although there may be federalism without a federation, there can be no federation without some matching variety of federalism', or as he better put it (p. 76), 'some federalism is always implicit in any given federation at any given time', sound ever more convincing. Thus, our discussion on Russia should examine not the *absence* of federalism in the societal and cultural sense of the word but rather its *nature* and the nature of its *relationship* with the federal structure or the federation. For the problem of federalism, as Livingston (1956, p. 9) observed, is to make 'instrumentalities fit the society beneath'.

In this connection, King (1982, p. 75; see also Stepan 1999, pp. 22–3) insists any contemporary federation can be consistent with any of at least three types of federal thinking. A federation can be established in order to secure greater centralisation, or with a view to greater decentralisation, or to achieve some form of balance.[2] The first – centralising – group of federations (examples being the USA, Canada, Switzerland, and Australia) includes countries with a history of previous existence as distinct colonies or states, which for a number of economic, security and strategic reasons decided to unite under a single independent government. Furthermore, these territorial units had sufficiently similar political ideologies and institutions, which created a solid foundation for unification (see Wheare 1963, pp. 44–9).

The centralising group of states did not have to work on the establishment of regional institutions. In turn, the primary objective was to create an effective national government. Implementing this goal required special policies such as the delegation of regional rights to the initially weak federal centre, which at the time did not threaten the territorial integrity of the composite units. In this case, the loss of sovereignty on the part of composite units represented the most sensitive matter. However, centralisation is generally a gradual process, which would see an increasing role for the national government over many decades.

The other group of states – Spain, Mexico, Belgium, Brazil, Venezuela, and India – have adopted federal constitutions because of the decentralisation of previously unitary states. In comparison with centralising systems, the situation in this case was entirely different and much more complicated. William Riker (1964, p. 11) for example went as far as to warn that such systems are likely to succumb to regional conflict or even civil war. Indeed, decentralising federations had already had a national government and had to work on establishing regional governments and on an appropriate way of devolving power to these governments without violating the territorial integrity of the existing state. This process is in itself very sensitive because it involves sharing central authority with the regions, which would naturally attempt to seize the maximum degree of power. Therefore, as Stepan (1999) argues, pursuit of these policies should rely on the creation of a system of economic and political incentives for regional governments to remain within the already existing state and to accept the centre's supremacy in certain spheres along with the binding effect of the constitution.

Having said this, such a strict division between the existing federal philosophies applies largely to the early stages of federal development. The subsequent dynamics demonstrate substantial changes in the normative orientation of federal relations. In broad terms, we are observing the interaction between structural and cultural patterns – a relationship that is widely regarded as being extremely intricate, intimate, and mutually reinforcing. The dynamic of this relationship depends on the matrix of social and institutional parameters, as well as on external contextual stimuli that substantiate their interdependent evolution to produce new and complex forms of political interaction (Alexander 2000, pp. 35–7; Welch 1993, p. 103; Burgess 1986, p. 22).

Analysis of this problem led many scholars (Eckstein 1998; Almond and Verba 1963, pp. 20–30; Brown and Gray 1977, p. 4; Greenstein 1969, p. 7) to conclude that there is never a complete fit between perception and reality, between abstract ideas and implemented practices, and more broadly, between societal processes and institutional structures that reflect them. Applying this logic to federations, King (1982, pp. 75–6), supported by Vile (1977, p. 2), insists that federations, as institutions, are in flux and will invariably 'betray some moral and normative orientation' (i.e. federalism). As we have already observed, he also insists that the normative orientation of the federal vector is subject to a similar change. Respective shifts within both federalisms and federations are bound to produce developments, which Eckstein (1998, p. 24) termed as 'adaptive change'. Indeed, federations' structural fluctuations could, in Watts' (1966, p. 16) words 'subsequently influence loyalties, feelings, and diversities', thus producing new dimensions of societal federalism and respectively highlighting either its integrative or decentralising facets. The evolution within federalisms, at the same time, could provoke some structural responses within federations.

I shall bring up the two simplest examples relevant to our case. If the federal structure shifts towards excessive centralisation and does not adequately reflect current socio-territorial cleavages, normative orientation of the real federal relations will invariably be headed towards diversity and creation of a looser union. This could take open forms of public and institutional pressures – like in the Fraser era Australia (Sawer 1977, pp. 18–19), the United States of the *new federalism* period (Vile 1977), and Canada. Such pressures could also assume disguised 'adaptive' forms. In this case, the requirements for autonomy find alternative venues and routes of expression (Nigeria, Brazil, and Venezuela during the gubernatorial appointment periods). If, on the other hand, an institutional network is not centralised enough to adequately reflect the realities of the existent federal process, the desires for a stronger unity may prevail and accelerate centralising tendencies. The cases in point are the European Union, or the United States, Brazil, and Switzerland during the early stages of their historical developments.

The early Yeltsin era witnessed a very emotional outbreak of decentralist federalism that gained momentum following the dissolution of the Soviet Union. From this point of view, Yeltsin's famous 1991 phrase 'take as much sovereignty as you can swallow' represented a federalising institutional response that was proposed to settle the bargaining conflict between the national centre and the component territorial communities. The resulting *federation*, however, had become excessively decentralised. Most observers tended towards a consensus that it was ineffective, fraught with separatism, and in the long run incompatible with the development and modernisation of Russia. The regional disregard for the constitution (Sakwa 2002; Stepan 2000; Kahn 2000; Petrov 2000a), the unclear division of powers, the alarming inequality between the regions in their relations with the centre (Solnick 1995; Solnick 1996), the absence of integrating political parties and networks of intergovernmental co-operation (Stoner-Weiss 2002; Ordeshook 1996; Golosov 2003) have been among the most commonly discussed themes. Many analysts denied this structure the name federal, binding it closer to a confederal (Hahn 2003, p. 115), or even more often *treaty-constitutional*, arrangement (Rumyantsev 1995; Plyais 1998; Kahn 2000; Lysenko 1998).

A strong case can be made that such a loose structure did not adequately reflect the objective realities of the existent federal processes, and in particular, their integrative aspects. Thus, serious political and societal forces had emerged towards the end of Yeltsin's rule that sought to render the Russian federal structure a more coherent centralising, and institutionally entrenched basis. This book will examine such issues in detail while discussing the initial conditions that led to the emergence and consolidation of Putin's federal reforms. I will argue that Putin was at once faced with the challenge of building an adequate institutional federation on the basis of the existing federal processes. Russia's second president was, therefore, tasked with creating a structure in which the supremacy of the federal law and constitution would be guaranteed, secession would be outlawed, and intergovernmental co-operation would flourish on the principles of reciprocity. At the same time, implementing these initiatives required some considerable consolidation of a political will by the federal centre. In this context, the key problem was to strike a careful balance between institutional reinforcement and provisions for the genuinely federal nature of centre-regional relations.

It soon became clear that the Kremlin was ignoring the existence of such a dilemma and taking the policy of state building to the extreme. Despite the launch of a number of modernising initiatives, the general vector of Putin's reforms was aimed at the re-centralisation of the Russian federation and the establishment of the political authority of the federal centre over the regions. In this case, the stylistic pattern of Russia's political culture somewhat marred the initially balancing intentions. This created a structure that yet again failed to reflect the existent federal realities, albeit from a centralising, rather than loose decentralising, Yeltsinite angle.

The discussion in this book will further show that, with the erection of this rigid framework, Russian federalism as an institutional process and state of society has not disappeared. And it could not have done so in such a short period of time within such a vast territory with its highly diverse ethnic and cultural society. In turn, this federalism has changed its integrative late Yeltsin-era vector and assumed a new decentralist orientation. It has also become adaptive in form and begun searching for an alternative means of expression with the aim of effecting devolutionist structural moves. The follow-up question emerges as to which direction Russia's federalism-federation alliance is headed. As I have already mentioned above, the growing incongruence between the structure and the process of Russia's federalism will ultimately move the system into a qualitatively new condition. The question remains as to whether the decentralist thrust of actual federal relations will eventually be able to effect institutional relaxation or whether the hierarchical structure will prevail and alter the normative orientation towards a centralist stance. While there are grounds for a range of optimistic thoughts, we should also warn of the 'cultural reversal' dangers that may subvert any potential institutional or societal moves towards liberalisation and devolution.

Potential devolution: an optimistic scenario

We have already discussed that the interaction between the structure and the process of a federal system depends on the complex interplay of institutional and social factors but also, and more importantly, on the existence and nature of the external stimuli. Thus, in the attempt to answer the question above, we need to account for a broad spectrum of economic and socio-political environmental conditions. It could be argued that Russia has already developed a range of social and economic factors that could coalesce with the decentralist stance of the federal relations and produce some serious pressures towards a devolution. From the economic point of view, Russia may follow two potential scenarios, both of which are conducive to structural relaxation.

In the optimistic case, Russia would be able to capitalise on favourable commodity prices and successfully divert her financial resources towards the high technology, agriculture, and consumer goods production sectors. The consequential development of urban infrastructure could soon raise pressures for a greater democratisation of political relations, even from a theoretical point of view (Lipset 1960; Przeworski *et al.* 2000, p. 50 and p. 111). Many such developments are already underway in light of Russia's current economic success. Moscow Institute of Sociology estimates that by 2010, the middle class, which espouses a liberal-conservative ideology,[3] will comprise some 35 per cent of the population.[4] It is also important that Russia's medium-sized and small businesses have already begun to actively finance various social movements, accounting for an almost 40 per cent share of the entire associational 'market' (Alexeeva 2007). If these dynamics persist, provision of a greater level of regional autonomy, self-governance, and participation will become absolutely indispensable.

In the pessimistic case, a potential decline in world energy prices could foment economic difficulties associated with the depletion of the state gold reserves and stabilisation fund. Russia's Institute of Transition Economies estimates that these funds could last just over two years should the oil price drop to its all-time average of £20 per barrel (Gaidar 2007). Given that the current federal construction is fundamentally based on the principles of economic determinism and subsidiarity, such a situation could compromise the Kremlin's ability to deliver the grants and lead the regions to reshape their dialogue with the Kremlin towards a greater decentralisation.

A range of social factors is also consistent with the existing pressures towards decentralisation. First, the newly elect President Medvedev will have to consolidate his existing political relationships in the regions in order to ensure effective governance. This would be impossible without a certain degree of compromise and devolution. It is also significant that Medvedev is well acquainted with regional developments because of his previous role as a chief of the national projects system. More importantly, Putin's further participation in politics as a Prime Minister invariably creates two potential centres of power. Such a situation is *a priori* unstable, despite Putin's legitimacy drawn from the *Edinaya Rossiya* (ER; United Russia) constitutional majority within the parliament. Thus, it is logical to assume that the Kremlin will navigate political corners smoothly avoiding any sharp conflicts. Maintaining this fragile construction will be impossible without striking productive relationships with the regional elites and granting the latter some form of autonomy.

Second, the development of a civil society is accelerating against the backdrop of the Kremlin's resistance. Genuine grass-roots associations – movements for motorists' rights, ecological, property holders, and professional movements – actively speak on practical socio-political and economic problems.[5] From a theoretical point of view, the emergence of such civic activity could effectively lead to a growth of what Putnam (1993, pp. 86–7) termed, social capital of the regions. Social capital, which refers to the ability of citizens to actively participate in independent associations and to take part in the social life of their community, is largely associated with the ability of the regions to defend their interests at the national level (Putnam 1993, pp. 98–9 and p. 105). I do not wish to claim that such organisations could bring an overnight change to the functioning of Russia's centre-regional and, in a broader sense, political system. However, these dynamics will certainly create considerable social pressures towards establishing a greater institutional reciprocity in the centre-regional political life. Thus, we can see that there is a range of factors in place that could indeed move Russia's federal structure and process in a more progressive direction.

Potential institutional problems and the political style factor

The success of this undertaking, however, remains rather unclear. While the current vector of Russia's federal relations is geared towards devolution, it is extremely premature to assert that the course of this process is protected from arbitrary, authoritarian swings. Indeed, the state's liberalising and democratic initiatives are in danger of not reaching their audience, while the process of real federal relations can already be seen as adaptive and non-transparent in form. Such problems stem from the existence of some serious constraints at a broader cultural level, which are considered to be traditional and authoritarian. Pye and Verba (1965, p. 520) contend that 'traditional beliefs may lead to fundamental modification of innovative institutions so that they fit the traditional culture'. Building upon this argument, Pye and Verba (1965, p. 5) warn that it is important to determine how 'democratic values and modern political institutions can be most readily transferred to new environments'.

Indeed, a similar range of liberalising measures implemented in differing socio-historic conditions is unlikely to produce identical outcomes. For cultural and socio-historical context matters, it puts the newly emerged ideas and structures through its own prism and shapes them to fit its framework and match its limitations. As Skinner (2002, p. 176) notes, 'there cannot be a history of unit ideas as such, but only a history of the various uses to which they have been put by different agents at different times'. The way in which such agents implement the newly emerged ideas and institutional practices is, undoubtedly, influenced by their 'attitudes towards the political system and its various parts and attitudes towards the role of the self in the system', (Almond and Verba 1963, p. 12), in short, by what is defined as political culture. Greenstein (1969, p. 7) succinctly observes that 'attitudes' are permanently engaged with the institutional environment in a 'push-pull relationship', in which the 'pushing' attitude determines how the actors operate a 'pulling' structure.

In transitional societies, and especially in Russia, where almost all democratic institutions have been externally borrowed, the interaction between political culture and institutions is expected to be very complex. In particular, the gap between the idealised and implemented institutions is likely to remain wide for a considerable period of time. For, as Brown (1977, p. 4) argues, the speed of institutional changes is much more rapid than that of cultural adaptation and it can, therefore, create, 'dissonance' or incongruence, between 'the political culture and political system'.

One important theme that binds the system of beliefs, values, and attitudes with institutional behaviour is that of *political style* (Putnam 1973, pp. 34–45; Waterman 1969, p.. 113–14; see also Spiro 1959). Verba (1959, p. 549) ascribes political style to the 'structure of formal properties of political belief systems that is not the substance of the beliefs but the *way* in which the beliefs are held'. He (1959, p. 549) further notes that *political style* 'lies on the border between the system of political culture and the system of political interaction, and involves those informal norms of political interaction that regulate the way in which fundamental political beliefs are applied to politics'. Thus, political style defines not *what* politicians believe but *how* they believe and *how* they interact politically in order to implement their ideas.

The *political style* concept was substantiated by the work of social psychologists, such as Rokeach (1960), Moscovici (1977, pp. 110–51), Adorno *et al.* (1950) and many others. Rokeach (1960, p. 7) in the influential *Open and Closed Mind* concludes that, if we know something about the *way* a person believes, it is possible to predict *how* he will go about solving problems that have nothing to do with his ideology. In particular, he (1960, p. 13) insists, 'authoritarianism and intolerance in belief and interpersonal relations are surely not a monopoly of Fascists, anti-Semites, Ku Klux Klanners, and conservatives. We have observed these phenomena ... among persons adhering to various positions along the total range of the political spectrum form left to right. We have observed them in religious circles; in the academic world where the main business at hand is the advancement of knowledge; in the fields of art and music, and so on'.

It now becomes clear that, while certain policies and institutions can be borrowed from various external sources, the outcomes of such borrowings will ultimately depend on the political style of the borrower. And, while some policies can come across as democratic in form, the *way*, in which they have been implemented, could be undemocratic, and even overtly authoritarian. Speaking of post-Communist Russian politics, Sakwa (2005, pp. 52–3) suggests that 'authoritarian and personalised *style* [my emphasis] of politics' represents one important trace of the old Soviet system. Many other authors also contend that, during various periods of Russia's history, most politicians, from the democratic spectrum or otherwise, pursued their policies in a predominantly authoritarian political style.

Timofeyev (2004, p. 93) insists that, despite having derived their views from Western neo-liberal notions, Russia's liberal intellectuals still wished to rely on the 'power of a strong state in order to promote the liberal programme among a reluctant population'. Brown (2005, p. 189), in support of Lukin (2000), claims that the post-Soviet Russian reformers, despite the fundamental changes in their political beliefs, retained their essentially authoritarian style of political conduct:

> The 'democrats' decisively rejected most of the ideological content of Marxism-Leninism but they retained an absolutism, an impatience, and a lack of concern with the institutional requirements of political pluralism that owed much to their upbringing in a longstanding, and indigenously established, Communist system.

This resembles the propensity of Russia's liberal thinkers of the nineteenth century to perceive the world in a structured dichotomous fashion, inadvertently stereotyping political situations and creating various moral dogmas. Berlin *et al.* (1978, p. xvii), while admitting certain moral dilemmas of the Russian thinkers of that period, nevertheless observe their tendency to take 'the ideas and concepts to their most extreme, even absurd, conclusions' and to 'discover some monolithic truth which would once and for all resolve the problems of moral conduct'.

Personification of power, which like authoritarianism has direct links with informal political conduct, represents another important stylistic trait distinguished by academic literature. The argument has often been given explicitly historical, and at times theological, dimensions. White (1977, pp. 29–30; see also Pipes 1974, p. 74) traces the historical foundation of this phenomenon. He points out that the Russian word for 'state' (*gosudarstvo*) derives from the word for 'lord' or 'ruler', and is used to signify a power exercised by such a ruler over a 'dominion' or 'patrimony'. Kharkhordin (2005, pp. 4–10), as well as Skinner (1989, pp. 164–7), further argues that, while in English and in Italian *the state* and *lo stato* began to represent an apparatus of government independent of both the rulers and the ruled in the early seventeenth and sixteenth centuries respectively, in Russian this separation took place only in the early to mid-eighteenth century. Moreover, the notion of the state in Russia was further developed around this time to introduce the idea of a 'common good' that ultimately reflected the merger between the state, the autocrat, and the society, including the civil society (Kharkhordin 2005, pp. 14–15).

Theological ways of explaining the personification of power phenomenon particularly concentrate on such a merger. Kharkhordin (2005), Berdyaev (1955), Petro (1995) and others demonstrate that Russia's traditional ecclesiastic ethics advocated the lack of separation between the state, religion, and civil society and thereby led to a personalised and informal style of political interaction. Kharkhordin (2005) provides a fascinating account, linking Russia's historic ecclesiastic ethics and her contemporary political dynamics. Following Berdyaev (1955), he contends (pp. 50–1) that the Russian Idea, as seen in the Dostoevsky project, essentially implied supplanting the secular state and its use of the means of violence through the deification of man, reconstruction of the world on church principles, and by 'bringing church means of influence to regulate all terrains of human life'. This ethical ecclesiastic culture led to the lack of separation between the state, religion, and the civil society, in direct contrast to the states, which followed Catholic and Protestant religious traditions. This, in turn, resulted in the prevalence of informal means of control over individual actions, and informal collectives, rather than states and rules to impose behaviour. Petro (1995), in an attempt to demonstrate some democratic shifts in the Russian political culture, still shows that the new Orthodox thinking has not fundamentally moved away from such a philosophy. Discussing the most recent currents of Russian Orthodox theoretical theology, he points out (1995, p. 156) that the most progressive post-Soviet Russian clergy has admitted the need to separate the State from the Church. However, the idea of a symbiotic union between the two, in which each entity is involved in a 'close and intimate relationship, while preserving the self-sufficiency and independence of each' remained prevalent. Thus, even the most recent ideas sought 'sanctification of all aspects of human life, including political life'. Gray (1979, pp. 255–6) pointed in a similar direction two decades earlier. He however lamented various methodological difficulties associated with the production of a systematic analysis that could underpin this hypothesis.

It now becomes clear that personification of power, propensity for taking ideas to their extremes, and arbitrary-authoritarian mode of governance represent the most indispensable traits of the Russian *political style*. It is also important that such traits have a long and extended history that is difficult, if impossible, to overcome. The resulting informality of political relations coupled with adaptive and disguised forms of exercising intergovernmental dialogue and bargaining could impede the development of open and transparent forms of federal interaction. On a broader scale, Brown (2005, p. 197) argues that the prevalence of the informal means of political conduct might cause serious problems for the consolidation of democracy. Potential critics of these propositions may point to my earlier argument on the emergence of decentralising shifts within the Russian society and to the findings of many authors (Gibson 2001; Colton and McFaul 2002; Whitefield 2005) that contemporary Russians are generally supportive of basic norms of democracy. However, as our discussion above has demonstrated, adherence to authoritarian and personalised *style* of politics is not inconsistent with the espousal of democratic beliefs. For *what* can be done to achieve these democratic (or any other, for that matter) ends is a different story.

The structure of the study

We can thus see that Russia is facing a dual transformation problem in the regional dimension. On the one hand, the existing societal realities of territorial and ethnic diversity restrict the overwhelming acceleration of centralising arbitrary trends. On the other hand, these democratising devolutionist moves could be impeded by cultural stylistic constraints and take on disguised adaptive forms. It is also important that a similar process follows the introduction of Western-inspired modernising institutions. The mode of their operation invariably follows a distinct stylistic pattern and thereby creates a rift between the emergent structure and its functioning process. The rest of this book brings into focus the ongoing interplay between the structure and the process of Russian federalism, as well as the influence of political cultural constraints on the dynamic of interrelationship between these two concepts.

Chapter 2 argues that the emergence of Putin's reforms was predetermined by a number of historical, legal-political, institutional and socio-political factors. The discussion demonstrates that the structural pillars of Yeltsin's federal model did not comply with the standards generally established for federations across the world. The regional model was extremely decentralised and often threatened the territorial integrity of the Russian state. This resulted in the emergence of political and societal forces that began to press for the introduction of centralising policy responses. Chapter 3 examines the new balance of power established between the centre and the regions in the wake of the federal reforms. It assesses which aspects of the regional system were subject to significant change and which remained intact. A study of the tactics and strategies employed by regional leaders and the federal centre in the attempt to defend their political positions within the hierarchy of the national elite will complement this discussion. The chapter shows that, despite the initial launch of hierarchical centralising measures, the reforms resulted merely in a redistribution of power between the two levels of government and not in a fundamental restructuring of Russia's federal system towards genuine federalism. The centre failed to achieve an overt regional compliance in many key policy areas, and the dialogue between the two parties has remained non-transparent, complex, and far from being easily controllable.

Chapter 4 investigates changes in the functioning of those central institutions responsible for decision-making in the sphere of centre-regional relations.

Reference is made to elements of the Russian institutional system and to developed federal democracies in the West, with the aim of identifying fundamental factors that ensure political stability in successful federations and assessing how far these were established in the Russian case. Much attention is paid to the idea that successful federations require not only effective institutional structures but also the commitments of participating actors to operate these networks in a manner consistent with the main tenets of federal ideology and culture. The discussion will suggest that, while institutional networks of centre-regional functional co-operation are gradually emerging, their operational logic lacks the necessary commitments to the letter and spirit of federalism. This further contributes to the widening gap between the structure and the process of Russia's federal model.

Chapter 5 builds upon this argument and examines major institutional developments in the regional sphere. It looks at the ways in which political and business interests are articulated in regional institutions and investigates changes in the composition of regional elites since the advent of the federal reforms. The discussion particularly focuses on those of the Kremlin's policies that aimed at modernisation of regional economic conditions and introduction of a greater transparency in the regional socio-economic climate. The role of Russia's political style in the implementation of these policies will be discussed as one significant factor impeding the fruition of progressive initiatives.

Chapter 6 examines the introduction of the system of gubernatorial appointments in Russia's regions and discusses the previous model of electing the governors by direct popular voting. The chapter analyses the unintended consequences of this centralising move and argues that the introduction of constrictive legislation inadvertently led to the emergence of decentralising trends. Here again we shall see the growing rift between institutional intentions and the real political processes that take place within the erected structures. Our discussion will show that in the longer term, such societal dynamics may cause more problems for the Kremlin than the stringent institutional innovations could resolve.

Chapter 7 outlines the nature of Russia's inter-regional integration and the problems pertaining to inter-ethnic relations and regional mergers. The discussion focuses on the central attempts to take control over the institutions of inter-regional co-operation and to conduct the policy of regional standardisation within the ethno-cultural dimension. While the Kremlin managed to erect a range of new hierarchical structures, it has ultimately failed to effect a complete unification of Russia's regional space and to placate the struggle of ethnic groups for a greater political recognition and self-determination.

Chapter 8 concludes that the implications and consequences of the existing rift between the structure and the process of Russia's political system could be rather profound and the future of the current model uncertain. In this light the extant federal system represents a transitional model that puts pressure on the new president to take further steps either towards authoritarianism and expansion of the influence of federal administrative elites or towards a self-imposed power restriction and the liberalisation of various aspects of political life. The first option cannot be implemented without violating existing constitutional limits, which would be tantamount to a radical reversal of the course already taken in economic and foreign policies. The pursuit of liberalisation and perseverance with further institutional reforms aimed at a greater political transparency and democratisation represents the only possible solution to the range of problems facing the development of Russia's federal relations.

2 Russia's federalism under Yeltsin

The structure, process, and origins of Putin's reforms

This chapter will examine the particularities of Russia's federal structures and processes that predetermined the emergence of Putin's centralising reforms. The principal argument is that some serious institutional deficiencies in Russia's federal framework prompted the emergence of corrective integrative processes within Russian society. In the structural sphere, the discussion would practically demand gauging the position of Yeltsin's Russia vis-à-vis some central aspects of federal theory and establishing to which definitional category of institutional arrangement Russia belonged in terms of compliance with accepted norms for existing federations. It will be demonstrated that the regional system seen under Yeltsin was incompatible with academically established criteria for the federal states. Moreover, the legal-political shortcomings that accumulated within this system carried the danger of latent separatism and in the long run could have had destructive consequences for the future development of Russia.

This state of affairs was unacceptable to Russia's existing federal society, which had begun to search for new ways and means of correcting the balance. Thus, the federal process obtained an overwhelmingly integrative vector, and a broad range of socio-political, ideological, and cultural dynamics sustained the emergence of centralising reforms. That is why the ideas behind Putin's federal programmes had strong continuity with the plans for reconstructing the regional system, which surfaced in the Yeltsin administration during the second half of his presidency. Yeltsin's officials searched for ways of enhancing the effectiveness of the presidential representatives in the regions, dismissing the regional governors, altering the method of formation of the Federation Council, and harmonising central and regional legislation. However, all attempts to reform the regional system 'from above' were unsuccessful due to the lack of political consolidation at the federal centre and its weakened authority. More importantly, the regions also displayed a great deal of determination to redefine their existing inter-relations with the centre. The ideas expressed by the heads of the regional administrations converged with those of the centre on many points. However, the lack of co-ordination in the regional movement left these plans unimplemented. The existence of such trends demonstrated that, by the end of the Yeltsin era, both the regions and the centre had made a number of tentative reciprocal steps towards establishing a new federal framework.

A range of cultural, stylistic factors also became prominent in the emergence of the federal reforms. The ability of the Russian political elite to opt for liberal solutions remained rather limited due to the long-standing tradition of authoritarianism and the established lenience towards the 'strong hand' governing style. These aspects largely predetermined the choice of centralising, constraining methods while dealing with difficult political situations. In addition, the discussion of path dependency should involve an analysis of Russia's historic tradition of oscillating between centralisation-devolution cycles. Such an 'oscillation' dates back to the days of the formation of the Russian state. This dynamic to some extent predetermined the advent of Putin's re-centralisation as an historically driven response to the devolution of the Yeltsin era. Finally, it is also significant that Putin's tactical manoeuvres and the political climate of spring–summer 2000 enabled the president to defeat any possible opposition to his federal ideas and persuade society to accept his reforms.

Russia's structural problems

Yeltsin's federal institutions developed as a symbiotic mixture between federation and distinctive elements of confederation, which led many analysts and politicians to a wide acceptance of a new definition to refer to this quasi-federal institutional structure. The term *treaty-constitutional* federation aimed to reflect the institutional arrangement in which contradictory centre-periphery bilateral agreements – which often violate the supremacy of the federal constitution – play an important role in regulating centre-regional relations.[1]

The emergence of this term was associated with the fact that, while generally attempting to comply with accepted federal models and theories (at least in its declared constitution and name), the Russian federal arrangement did not conform with the two most important federal criteria – the supremacy of the federal constitution and the existence of a clearly delineated division of powers between the regions and the federal centre. The federal constitution has not succeeded in becoming a supreme and binding document for all the subjects of the federation; and various centre-periphery treaties, which created asymmetry between different regions in their relation with the centre, marred the principle of the division of powers. These deficiencies in the federal arrangement raised questions of stability and longevity of the treaty-constitutional system and by doing so created the legal and political basis for the launch of federal reforms with the advent of Putin.

Arguably, the violation of the principle of the supremacy of the federal constitution and the associated legal-constitutional and political contradictions stemmed from the country's geopolitical problems of an historical and geographical nature. The head of Yeltsin's committee on the division of powers between the federal centre and the subjects of the federation Sergei Shakhrai (1994) argued that the geopolitical component of Russian federalism has always rested on the virtually irresolvable dilemma between ethno-territorial and administrative-territorial models of federal relations.

Russia has a majority of ethnic Russian regions, which collectively comprise four-fifths of its territory, and a remaining number of ethno-territorial formations, in which popular pressures to safeguard national cultural and historical sovereignty are quite high and legitimate (Khakimov 1998).[2] At the pinnacle of these structures since December 1993 has been the constitution of the Russian Federation, which in accordance with classical definitions should represent one of the main prerequisites for the establishment of a genuine federation and must be 'supreme and binding upon general and regional governments'. Reconciling the all-inclusive nature of this document with the process of rapid sovereignisation and the extreme growth of national self-determination that followed the dissolution of the Soviet Union and democratisation of Russian society represented the core of Russia's problem.

For example, the leaders of ethno-territorial units often suggested that by virtue of representing the ethnic non-Russian minority their regions should be entitled to a 'special constitutional status' and to differing rights that 'would take into account the specificity of regions' indigenous populations'.[3] This became a political trend among many of the leaders of ethnic republics during Yeltsin's era. Solnick (1996, p. 22) characteristically cites a speech by President Murtaza Rakhimov of Bashkortostan, in which he makes the following statement: 'There are those among us who want to make the republics, oblasts, and krais completely equal politically. That cannot be allowed. Economically, they must be all identical... But there are questions that arise, for instance, in Bashkortostan and Tatarstan, that do not arise in the oblasts'.

This logic served to elevate many of these regions legally and politically above the centre and ultimately led to numerous violations of the 1993 federal constitution: at the end of the 1990s, more than 3,500 regional laws contradicted the country's basic law.[4] The constitutions of many republics superseded federal law and contained clauses on state sovereignty, citizenship, and limited foreign policy – which in itself denied the sovereignty of the Russian Federation as a whole and thus rendered Russia's institutional arrangement confederal. For example, in the republics of Bashkortostan, Kabardino-Balkariya, Yakutiya, Komi, Adygeya, Dagestan, and Tyva, the republican constitutions were considered more important than the federal basic law.

Articles (70) and (1) of the constitutions of the republics of Bashkortostan and Tyva respectively declared that their membership of the Russian Federation is based not on the federal constitution but on the Federal Agreement and existing bilateral treaties (Lysenko 1998, p. 32). Several republics such as Adygeya (Article 56) and Dagestan (Articles 1 and 65) reserved the right to suspend federal legislation and the constitutions of Adygeya, Buryatiya, Ingushetiya, and Kalmykiya stipulated the right to introduce an emergency (military) regime.[5] Moreover, the republics of Tatarstan, Bashkortostan and Komi established a dominant role for particular ethnic groups in their constitutions. Article 108 of the constitution of Tatarstan required that the president of the Republic must know both the Russian and Tatar languages virtually ensuring that only a Tatar could be elected to the office in the near future (Gorenburg 1999, p. 262; see also

Gorenburg 2001). In addition, Tatarstan, Bashkortostan and Sakha Yakutiya pressed for a change of passports for the regions' residents and insisted on the necessity of introducing a nationality column in these documents.[6]

It is clear that by claiming entitlement to a 'special status', the extreme independence of the regional laws, and other privileges, the regions under discussion encroached upon the constitutional rights of other subjects of the federation, as well as on the supremacy of the federal constitution and its binding effect on all. This situation also contradicts Articles 5.1 and 5.4 of the constitution, which both state that 'republics, territories, regions, federal cities, and autonomous regions and autonomous areas shall be equal subjects of the Russian Federation' and 'shall be equal among themselves in relations with the federal bodies of state power'.[7]

The problem, however, did not only end in the anti-constitutional claims of regions and ethno-territorial units. Rather, as Vladimir Lysenko (1998, pp. 34–5; see also Kahn 2002) insists, the federal constitution itself also contained a large number of serious contradictions that engendered regional non-compliance with its provisions. For example, the document does not outline the principles of including regional legislation in the unified federal legal system and mechanisms of federal interference in regional legislation when that legislation is in violation of the federal constitution.

In addition, by granting a different name or status to each group of administrative units and by permitting change in a subject's status through constitutional amendment 'and mutual consent of the Russian Federation and the subject of the Russian Federation',[8] Article 5.2 of the Russian constitution implies asymmetry between the federation's subjects. Moreover, such contradictions were inevitable from a historical-chronological point of view. The federal constitution came into force only in December 1993, after many republics had declared state sovereignty, enacted republican constitutions – as in the case of Tatarstan – and after the Federation Treaty came into force on 26 March 1992.[9] This situation implied the chronological supremacy of regional legislation and predetermined discrepancies between regional and federal legislation, as it was impossible to draft the federal constitution in a manner that would incorporate all existing regional constitutions and laws (Lysenko 1998, p. 34).

The deficiencies of Yeltsin's federal design in the sphere of centre-regional division of powers were mainly associated with the process of the conclusion of bilateral treaties between the regions and the centre. The federal constitution contains obscure provisions that permit bargaining between the regions and the centre, namely in Articles 11 and 76.2, both of which state that 'other regulatory legal acts can be adopted'[10] to formalise centre-periphery relations. The system of bilateral agreements that filled this constitutional niche began to substitute for basic law and resulted in legal and institutional inequality between the subjects of the federation in their relations with the centre in the sphere of division of powers.

Solnick (1996, p. 23) has rightly warned that 'these bilateral negotiations may ultimately supersede any constitutional norms of power sharing in Russia, provided that the centre abides by the terms of its deals. If it does, however, the Russian constitution might soon give way to a patchwork of ad hoc agreements – both published and unpublished – that are constantly undergoing a process of renegotiation'. It becomes clear that bilateral treaties violated the principle of division of powers between the centre and the regions by formally establishing the system of varying power sharing principles for each signatory region.

The need to establish the treaties at the heart of centre-regional legal dialogue emanated partly from the fact that the Russian legal system was not fully formed and was therefore insufficient. A number of essential laws were simply absent – for example, the principles of division of property between the regions and the centre were not defined, nor were the actual mechanisms of power sharing in the appointment policy for joint jurisdiction institutions. Licensing for the usage of natural resources, water, and land was under the mutual centre-regional respons- ibility, which opened the way to administrative chaos and corruption. Taxation, which under Article 72 (i) of the constitution falls under the mutual sphere of responsibilities, was not clearly delineated either. The appointment policy within important institutions such as notary, courts, and police force was not clearly outlined, officially falling under the sphere of joint centre-regional responsibility. In practice, the mechanisms of control and enforcement were absent, which opened the way to free interpretation of laws, regional arbitrariness, and admin- istrative manipulation.

In this case, the treaties began to substitute for gaps in the legislation and the Russian federation began to rely on a case-by-case agreement process between the centre and the regions; a process through which the regions attempted to manipulate the centre into conceding the maximum authority. In other words, the bilateral agreements cemented the trend of developing a dubious treaty-based process of centre-periphery relations instead of a legal-constitutional one. It is within this context that a number of experts (see Lysenko 1998, pp. 34–6) have noted that politicians at the time should first have established various provisions of the bilateral treaties as constitutional law before proceeding to enshrine them in negotiated agreements.

Naturally, the absence of appropriate federal legislation providing some ade- quate power sharing mechanisms between the regions and the centre and the gradual substitution of these mechanisms by an ever-growing number of bilat- eral treaties forced the governors to create their own system of sharing rights with the federal centre (Lysenko 1998, pp. 22–3). Article 72 of the federal con- stitution outlined spheres of joint jurisdiction between the regions and the centre, leaving the remaining areas to regional responsibility. However, no further offi- cial legislation emerged outlining how to implement this division of responsibil- ities in practice. Partly because of this situation, the judicial system has failed to enforce federal law across the land.

For example, many republican executives openly took control of appoint- ments in legal enforcement agencies, which Article 72.1(k) clearly defines as being under the 'joint jurisdiction of the Russian Federation and the subjects of the Russian Federation'.[11] Each region mastered its own way of co-ordinating appointments with the federal centre. Regional leaders effectively dominated regional branches of legal enforcement agencies: the presidents of the republics of Ingushetiya, Tatarstan, and Bashkortostan had the full power of appointment in these institutions (Pechenev 2001, p. 67).

More than half of the subjects of the federation created under the leadership of the local administrations' own Security Councils comprised the heads of federal tax police divisions, Federal Customs Committees, MVD (the Ministry of Interior), FSB (Federal Security Service) and prosecutors offices (Pechenev 2001). Because of this, local offices of the MVD, army and FSB operated within the system of vertical and horizontal hierarchy with dual responsibility to regional governments and to the federal centre. The heads of these agencies could not openly oppose central policies, but at the same time they had to be loyal to the governors, who financed the day-to-day needs of institutions' middle and lower ranking employees.

Within the framework of this discussion, the economic dimension of the problem deserves special mention. In his first address to the Federal Assembly on 8 July 2000, Putin insisted that legal inconsistency in the division of powers and violation of federal legal code spilled over to the economic sphere. Some regions not only enjoyed inexplicable privileges in servicing their fiscal duties, but also 'created difficulties for the free movement of goods and capital, introduced bans on trading grain and alcohol, and impeded the activities of some national banks on regional territory'.[12]

Of particular importance was regional independence in the appointment policy of federal economic agencies. This was mainly manifest in the problem of tax collection. Monopolisation of the State Tax Service (STS) branches by the regional administrations and reliance of the STS employees on regional budgets for housing, wages, and benefits resulted in a situation in which many regions managed to avoid their fiscal duties. For example, oil-producing Khanty-Mansi Autonomous Okrug, where the regional administration contributed some 260 billion roubles to the local STS branch, had the largest nominal tax arrears to the federal budget in 1996 (Shleifer and Treisman 2000, p. 135).

Moreover, because the law on division of property between the centre and the regions lagged behind the discussion of these issues in bilateral treaties, some regions – Sverdlovsk Oblast, Krasnodar Krai and Tatarstan being notable examples – concluded bilateral treaties defining the division of all regional wealth on federal, regional and joint levels without relying on any federal criteria in this sphere. This resulted in the transfer of many large regional enterprises to regional authority, which provided local elites with the opportunity to staff these structures with loyal managers and to withhold substantial tax revenues from the federal government (Lysenko 1998, p. 23).

It is clear that, as Yeltsin's era progressed, regional aspirations for economic and political self-sufficiency ultimately led to the country's poor economic performance, a weakening of the central state, and ineffectiveness of the fiscal system. Generally, a region's political independence determined its powers in the financial and fiscal spheres. Despite their wealth in natural resources, ethnic republics transferred a far smaller contribution to the federal budget than non-ethnic regions. In 1996, for example, Tatarstan transferred to the federal budget only 18.9 per cent of all collected taxes and Bashkortostan some 27 per cent. These figures were dwarfed by Moscow's contribution – as much as 57.6 per cent (Borodai 2002).

The discussion above demonstrated that Russia – though aspiring to establish a federal institutional framework at least in terms of its name and some constitutional provisions – failed to comply with established federal standards on the two most important criteria: supremacy of the federal constitution and appropriate legally established division of powers. It is clear that due to these factors the discussed treaty-constitutional arrangement contained strong elements of confederation. Indeed, the federal constitution did not have a 'supreme and binding' effect. Federal government did not operate directly upon people within many regions and various policy realms. Regional statutes were superior to the federal legislation in a vast number of territories. Needless to say, the auxiliary structures of the federation were either absent or seriously malfunctioning. Many academics and politicians at this point emphasised the destabilising impact of the treaty-constitutional system on the country's federal development. Kahn (2000, p. 82), for example, insisted that 'a hierarchy implicit in the treaty-constitutional approach raised republics above the federal government in all matters... From the point of view of the republics, in contrast to accepted federal theory, the Russian Federation was not greater than the sum of its parts'. These conclusions make it clear that the political climate in the country had reached the point at which reform of the federal system represented the most pressing necessity.

Ultimately, it appears that Russia faced three choices: (1) to maintain the current asymmetrical legal-constitutional arrangement; (2) to create a 'constitutional law for all' by raising all subjects' status to republican level; or (3) to enforce existing 'constitutional law' by taking contradictory privileges away from some republics and regions.[13] Maintaining an asymmetrical legal-constitutional arrangement is a democratic means of keeping ethno-territorial zones within the state's jurisdiction.[14] However, given that the constitutions of states with ethno-territorial asymmetries normally permit continuous bargaining between the centre and the regions, there must be a sufficient legal basis and strict rules for such processes, which incorporate both houses of parliament, the Constitutional Court, and the public. Ultimately, the 'quality' of an ethno-territorial federation with treaty-constitutional elements depends on the nature and dynamic of such bargaining processes between the centre and the regions. This answers the question of why some ethno-territorially formed countries successfully manage to maintain an asymmetrical legal-constitutional arrangement without violating major theoretical principles established for federations.

Stepan (2000) thoroughly examined the existing mechanisms of centre-periphery bargaining processes in order to set aside certain academic criteria and requirements for such states. He (2000, pp. 141–5) brings bilateral treaties into the focus of attention and advances the argument that the nature of the agreements plays a large role in defining the normative characteristics of ethno-territorial federations. Depending on their openness to political and civil society as well as compliance with constitutional law, these agreements fall into *constitutional, extra-constitutional* and *anti-constitutional* categories.

The first group does not raise any difficulties, as it comprises provisions envisaged by federal legislation and brought into the focus of public discussion following consensus. The second two groups, however, deserve special attention. These emerge in circumstances when leaders of ethno-territorial units in search of wider powers enter into backstairs political contracts with the centre, while publicly playing the national card by suggesting that their regions are entitled to wider rights and that these rights might in some cases contradict the acting constitution.

It is within this context that *extra-constitutional* arrangements relate to legal acts negotiated privately by the senior executive of the central government and the regional governor without bringing the matter into a forum for open political and public discussion. *Anti-constitutional* agreements represent legal documents of a fundamentally similar nature although some of their statements contradict the acting federal constitution. It is clear that the proportion of the last two groups of agreements in a country's legislation define whether it can qualify for definition as a legally bound democratic federal state.

Stepan (2000) illustrates the relative success of Spain where iterative bargaining between the centre and ethno-territorial regions of the Basque Country and Catalonia remains within democratic limits because of the absence of extra-constitutional and anti-constitutional treaties: agreements between the centre and the regions in Spain are made public, approved by both houses of parliament and the country's constitutional court. This system virtually precludes the possibility of excessively unjust rights for ethno-territorial regions vis-à-vis the rest of the federation and forbids the possibility of secession.

Conditions in Russia in 1999–2000 were radically different to the Spanish experience, which effectively precluded the possibility of accommodating successfully legal-constitutional asymmetry in the existing framework of federal relations. Indeed, as we have already discussed, establishing democratically orientated bargaining mechanisms between the centre and the regions faced the constraints of inefficient legal code and a number of chronological-formative problems. Russia's bilateral treaties not only contradicted the constitution and violated the principles of a division of power – and therefore were of an overwhelmingly extra-constitutional and anti-constitutional nature.

Stepan (2000, p. 144) insists that the vast majority of bilateral treaties that were negotiated and signed by the Chief Executive of Russia and the Chief Executive of one of the 89 constituent members of the Russian Federation were non-constitutional and were concluded without being signed by, or even shown to, the Russian parliament. Maintaining this overly asymmetrical legal arrangement, in particular the parts emerging from extra-constitutionally and anti-constitutionally negotiated deals, would represent an extremely dangerous decision for the entire federal structure. As this discussion has demonstrated, this trend was already leading the country toward constitutional collapse. Clearly, this was an unsuitable solution for a nascent Russian federation with its numerous socio-political and institutional problems.

The second option, establishing the supreme 'constitution for all' by raising the rights of all territories to republican levels represents an equitable solution – but only theoretically. Such policy could result in the imminent confederalisation and territorial fragmentation of the country because – as this chapter has discussed – many ethno-territorial units already exist within the federation on a confederal basis. The process of latent 'republicanisation' of Russia's regions underpins these concerns. For example, in March 1992 four autonomous oblasts – Adygeya, Gorno-Altai, Karachaevo-Cherkessiya, and Khakasiya – put pressure on the centre by demanding their constitutional elevation to republican status. By 1995, following the lead of these areas, some non-ethnic regions attempted to organise movements to advocate increasing the rights of the rest of the regions to republican levels. Sverdlovsk Oblast, Krasnodar Krai, and Orenburg Oblast demanded the conclusion of bilateral treaties with the centre allowing them fiscal and political privileges on the same footing as those of the republics.[15] Admittedly, the campaign's underlying idea relied on the assumption that Russia must not base its centre-periphery relations on an ethnic and national basis.

In practice, however, the proposed treaty between the centre and Sverdlovsk Oblast contained a large number of confederal elements, which overall could destabilise the country's institutional structure. These were modelled on the unsuccessful 1993 constitution of the Ural Republic, which aspired in some of its provisions to suspend federal legislation and take control over appointment policy in federal institutions operating on the territory of the republic (Lysenko 1998, pp. 22–3). The demands of Sverdlovsk Oblast fomented significant political instability, demonstrating that there was a real threat of confederalisation of the Russian Federation if the proposals were carried through. It now becomes apparent that raising other regions' constitutions to republican levels would lead – as in the first case – to a loose confederal arrangement between the existing Russian regions and threaten the territorial integrity of the current state.

The option of abolishing anti-constitutional rights from the republics could result in popular discontent and by doing so would encroach upon the voluntary (i.e. democratic) basis of the federation. However, if alteration of regions' anti-constitutional rights did not involve extreme solutions, it is possible to argue that this option could have been the most feasible way to conclude the matter. In his comparative study *Federalism and Federation*, King (1982, p. 60) states that 'for a federation to have a coherent procedure which makes it possible for final decisions to be taken in any sphere, it must be possible somehow to amend any or all of its rules. Therefore, if one of these rules empowers a locality to act in some area without reference to the centre, it will normally hold that the rule can, nonetheless, be overruled without requiring the consent of the unit against which such action might be taken'. It is also important that the legal harmonisation is a continuous process that takes place even in the most advanced democratic federations with the enactment of new regional and federal legislations. Given these theoretical stipulations and a difficult choice between the existing options, the Russian authorities opted for the third, constraining solution.

Russia's federal processes: the historic dimension

The implementation of such constraining measures was possible because the vast range of structural problems discussed earlier has led to the emergence of corrective centralising trends within Russia's political society and general public. Besides the socio-political dimension, however, this corrective process was also rooted in a range of historic-cultural factors. These factors view the evolution of Russia's federal relations in a cyclical pattern and see the advent of Putin's reforms as an historically driven response to the extreme decentralisation of the 1990s. Let us first elaborate on these historical factors and turn to the socio-political aspects in the following subsection.

Many analysts of Russian federalism – among them Arinin and Marchenko (1999), Arinin (1999a and 1999b), Gelman (2002), Petrov (2000b), Alexeev (1999) and others – have observed that the most important historical component of Russian federalism includes the fact that Russia's centre-periphery relations have always represented an oscillating process of centralisation-devolution taking place within a rigidly authoritarian unitary framework relying in the Soviet stage of its existence on an official ideology.[16] Indeed, a unitary state since the fifteenth century, Russia was composed of large divisions of provincial or local administrations, which were decentralised in practice and even demonstrated occasional separatist tendencies. The centre had to react to various autonomy pressures from these units and to conduct differing regional reforms in order either to improve governance or to display lenience towards greater liberalism. At the same time, such concessions were short-lived, as the promises for devolution or other regional liberalisation initiatives were conducted on a non-contractual relational basis and relied in its implementation on Russia's informal style of political conduct. In this context, the centre quickly betrayed its initial liberalising intentions and reverted the structure and practice of the regional dialogue back to the well developed unitary path. More importantly, such cycles were no minor fluctuations incapable of reversing the general vector of political development, similar to what we see in the nature of centre-regional relationships in the West. Rather, these were important policy milestones that seriously affected the evolution of independent regional initiative. Let us now take a brief look at the history of such fluctuations.

The regional reforms of 1550 launched by Ivan IV were quickly followed by a significant structural reversal that consolidated Russia's path towards autocracy. The initial policy ends, which were aimed at enhancing local self-government based on the emerging estate (*soslovie*) system (Pavlov 2003, pp. 73–5), were substituted by the introduction of the *oprichnina* policy. Peter's initial reforms of provincial administrations instituted ten *guberniya*s headed by appointed governors and established the elected local nobility councils as advisory bodies to these governors (Robbins 1987, pp. 6–7). However, Peter was quickly disenchanted with the decentralised form of government and transferred many gubernatorial responsibilities to representatives of central colleges, local military commanders, and the courts. Catherine's first moves towards instituting local self-governance were followed by the introduction of central ministries in 1802 that constrained even the pre-existing local initiative and intended to co-ordinate the functioning of the governors. The introduction of the *zemstvo* insti-tutions during the Great Reforms of 1864 was mitigated by some significant restrictions on their functioning. The counter-reform of the late 1880s and early 1890s provided appointed governors with increased leverage vis-à-vis the local *zemstvos* (Robbins 1987, p. 18; Conroy 1998, pp. 66–9).

The Soviet government also pursued a policy of cyclical centralisation-devolution development. The state centralised for the implementation of major economic tasks, such as industrialisation and collectivisation, and devolved power during brief periods of liberalisation – examples being the New Economic Policy, *sovnarkhozy*, and *perestroika*. Interestingly, Russian sociologists Arinin and Marchenko (1999, p. 77) have observed that the Soviet phases of centralisa-tion and devolution followed each other within a period of 35 years (1922, 1957, and 1992), as though indicating the capacity of central government to sustain devolution within certain unitary limits.

From this point of view, the establishment of the Russian Federation in the early 1990s represented – with certain reservations – yet another phase of decen-tralisation. This process was gradual and took place over a period of five to six years. The Chairman of the State Duma Committee on Regional Policy and Fed-eration Affairs Lysenko (1998, pp. 16–17) perceives five major stages of decen-tralisation in the 1990s. First, the starting point was the signing of the Federal Agreement on 31 March 1992. Second, the December 1993 Constitution of the Russian Federation raised the status of subjects of the federation to republican level and allowed the conclusion of bilateral agreements between the centre and the regions. Third, the centre signed the first agreement with Tatarstan in Febru-ary 1994. Fourth, Kabardino-Balkariya and Bashkortostan followed Tatarstan's example and the signing of treaties entered into a new popular stage. Finally, from 1996 the federal government began to conclude treaties not only with the republics but also with ethnically Russian territories such as Orenburg, Sverd-lovsk and Kaliningrad Oblasts, and Krasnodar Krai.

By 1997, some regional analysts (see Khakimov 1998, p. 41) had begun to speak of a complete victory of the decentralising tendency over unitary tradi-tions. In accordance with the logic of Russia's traditional centralisation-devolution oscillation, it was clear that centre-periphery relations had reached the point at which a centralising turnaround was imminent. This turn occurred with the election of a new president in March 2000.

Moreover, it is important not to overlook the fact that all the political proc-esses taking place within the newly created Russian Federation were in their essence different from the rest of the country's political experiences. With the removal of the authoritarian unitary framework, the government had to find a system of voluntary mechanisms with which to keep the state's constituent units together. This meant finding ways of establishing an appropriate way of delegating powers to regional governments without violating the territorial integrity of the existing state. As Alexeev (2000) notes, 'the absence of a unifying state ideology in post-Soviet Russia enabled the central government to provide the regions with selective economic and political incentives to remain part of Russia – a strategy that was unavailable to Soviet leaders in the context of institutions based on Marxist-Leninist ideology prescribing highly centralised one-party government'.

Following this logic, it is important to remember that emerging re-centralisation represented a qualitatively new phenomenon in contrast to the centralising trends of previous eras, in the sense that it took place within the federal framework of a democratising state. This leads to the suggestion that relating recent efforts at centralisation solely to Putin's 'reform from above' would be insufficient. Rather, existing socio-political forces permitted the implementation of Putin's policies. We should now move on to the analysis of the existence of reforming impulses and centripetal incentives on the part of regional political and business elites, which led to mutual tectonic movements towards re-integration.

Ideological and socio-political shifts of the late Yeltsin era

Analysis of central and regional political developments after 1997 demonstrates that the regions were generally dissatisfied with central policy while the centre could not find an optimal way of delivering sound changes. Therefore, expectations of re-centralisation emerged among both regional and central political elites as well as among the general electorate. These hopes were deeply rooted in the political developments of the late Yeltsin era and formed a significant socio-institutional basis for the emergence and implementation of Putin's federal reform.

Among the main socio-political factors that permitted re-centralisation the following three seem to be the most important: (1) the emergence of tectonic reciprocal movements by regional and federal politicians to redefine the existing framework of their relations, (2) the centre's weakness and growing attempts by the regions to take the lead in reforming the federal system, and (3) the emergence of the OVR (Fatherland-All-Russia) movement and its relation to the Kremlin's view on federal policy.

First, by the end of the 1990s, many analysts observed the existence of reciprocal movements between the central and regional political elites towards redefining the framework of their interrelationship. Researchers from the Moscow Carnegie Centre (Petrov and Titkov 1999; see also Petrov 1999), for example, insisted that the interactive political dynamic between the two given poles during the late 1990s was centripetal and called this general trend *Moscow-centric*. A number of important developments confirm these assumptions.

The phenomenon of vertical mobility among regional and federal politicians heralded the emergence of trends of centre-regional re-integration. Regional leaders aimed to secure key positions at the centre, while federal politicians attempted to seize influential regional posts. For example, Vladimir Zhirinovskii ran for gubernatorial elections in Belgorod Oblast in May 1999 and Gennadii Seleznev, the Speaker of the State Duma, was among the candidates in the gubernatorial elections in Moscow Oblast in January 2000. On the other hand, Boris Nemtsov and Sergei Kirienko of Nizhnii Novgorod, Vladimir Putin from St. Petersburg, the Mayor of Samara Oleg Sysuev, and the governor of Leningrad Oblast Vadim Gustov came to work in the federal government, and the presidential ambitions of the Mayor of Moscow Yurii Luzhkov were well known (see Chirikova and Lapina 1999, pp. 160–1). The old Federation Council also fulfilled the function of a springboard into federal politics for many regional politicians.

Moreover, regional leaders often capitalised on the centre's authority and prestige in the resolution of certain regional conflicts or in the development of various regional events. Examples of these tactics include gubernatorial elections in Krasnoyarsk Krai in May 1998, which evoked wide national and international reaction,[17] the March 2000 legislative elections in St. Petersburg that fomented controversy due to their timing, as well as mayoral elections in Nizhnii Novgorod in March 1998 and presidential conflicts with the opposition in Kalmykiya in October 1998. Kalmykiya represented a particularly interesting example. Following a financial conflict with Russia's Central Bank, which imposed sanctions on the National Bank of Kalmykiya for financial misconduct and corruption, the republican President Kirsan Illyumzhinov threatened his region's secession from the Russian Federation. The centre conceded to such 'reasoning' and reinstated financial flows to the republic.[18]

Furthermore, as Petrov and Titkov (1999) note, a number of regional governors were often speculating on federal political developments for the sake of future career prospects in federal institutions, and thereby national rather than regional political affairs predetermined their tactics and decision-making. These leaders normally manifested such a strategy by taking an active part in various nationwide political developments. For example, the President of Tatarstan Mintimer Shaimiev, the Mayor of St. Petersburg Vladimir Yakovlev, and the Mayor of Moscow Yurii Luzhkov were the founding fathers of the nationwide political movement *Otechestvo-Vsya Rossiya*, the goal of which was to propel the Mayor of Moscow to the presidency. The parliament of Tatarstan had passed an overly liberal (when compared with Russian national standards) land law allowing the sale of arable land – a move partly aimed to irritate the State Duma and please the Kremlin. The regional legislature of Kaliningrad Oblast echoed this initiative but failed to pass a similar document successfully (Petrov and Titkov 1999).

The phenomenon of centripetal developmental dynamic undoubtedly evokes a number of questions. Why, for example, did republican leaders, who fulfilled nationalistic claims for the regions' indigenous populations and obtained significant independence in their dialogue with the centre, have a high stake in becoming politicians of national prominence? Why, instead of pressing for further economic and political separatism, did regional leaders adopt a visibly centripetal stance? Answers to these questions are likely to be ethnically related as well as traditionally bound.

In the first case, the national composition of ethnic republics did not favour titular nationalities – a factor that greatly diminished the scope for further capitalisation on the 'national card'. Figures show that the ethnic composition of the existing national republics was highly intermixed: indigenous populations were not always in a clear majority, and territorial redistribution of nationalities did not correspond to the geographic boundaries of the existing republics. For example, a number of nationalities reside outside the republican or territorial borders. Tatarstan, Adygeya, Komi, and Bashkortostan are prime examples. In many republics, the number of Russians equalled the indigenous population and in some areas the Russian population was much greater. The overall number of nationalities and ethnicities residing in the republican territories varied between 50 and 150. Table 2.1 illustrates this in detail.

Arguably, a similar situation occurred in the Soviet Union. Rogers Brubaker (1996, p. 36) points out, for example, that 17 per cent of all Russians, one-third of Armenians and nearly three-quarters of Tatars were living outside designated republican borders. However, a broad range of circumstances – political, institutional, economic and historical – permitted the USSR's republican leaders not to pay much attention to their areas' national composition and pursue regional and ethnic separatism up to their secession. The set of variables changed within the circumstances of a semi-democratic state, which Russia had become by the end of the 1990s. Similar factors vis-à-vis national composition could oblige regional elites to take into account the sentiments of the non-indigenous population.

Furthermore, despite the growing autonomy of the regions in the economic and political spheres, the centre still had strong traditional authority as the financial and political backbone of the country. Chirikova and Lapina (1999, pp. 166–70; see also Chirikova and Lapina 2001) observe that the federal centre exerted strong influence over the Central Bank, Ministry of Finance, customs and other institutions central to regional economic activity. In addition, the centre had ownership of transportation systems and energy supplies; solid political contracts between the Kremlin and energy giants *RAO UES* (the country's electricity monopoly), *Gazprom* (Russia's gas giant), and the transportation monopoly *RAO RZhD* (Russia's railroads) transformed this ownership into a powerful instrument of political manipulation.

Clearly, many of the resource-rich regions – both ethnic and Russian – were heavily dependent on the centre's provision of these services. This factor made the leaders of these regions more lenient towards the centre and more interested in involvement in all-federal politics. Therefore, for many regional politicians rising up to the national level was not only a matter of prestige but also an essential tool of influence over central policy towards their respective regions. For example, 39 out of 56 deputies in Buryatiya's legislature – the People's Khural – voted for former general prosecutor Yurii Skuratov to represent the republic in the Federation Council. The deputies explained their choice by the desire to have a powerful person in the centre, whose strong connections and influence could

Table 2.1 Ethnic composition of a number of Russia's republics and regions[19]

Region	Population	Ethnic composition			
		Number of nationalities	*Percentage of indigenous population*	*Percentage of Russian population*	*Percentage of other major nationalities*
Bashkortostan	4,000,000	100	Bashkir 22%	39%	Tatar 28% Chuvash n/a Mari n/a Ukrainian n/a German n/a
Republic of Chuvashiya	1,360,900	More than 50	Chuvash 67.8%	26.7%	Tatar 2.7% Mordva 1.4% Ukrainian 0.5% Mari 0.3% Belarusian 0.1%
Republic of North Osetiya	676,922	96	Osetin 57.43%	24.81%	Armenian 2.2% Georgian 1.75% Ukrainian 1.2% Azeri 0.3% Tatar 0.27%
Khanty-Mansi Autonomous Okrug	1,401,900	123	Khanty 0.9% Mansi 0.5% Nenets 0.1% Total: 1.5%	66.3%	Ukrainian 11.6% Tatar 7.6% Bashkir 2.4% Others 10.6%
Republic of Kareliya	766,400	n/a	Karel 10%	73.6%	Belarusian 7% Ukrainians 3.6% Finnish 2.3% Veps 0.8%
Republic of Khakasiya	584,600	Over 90	Turkic 11%	79.5%	Ukrainian 2.9% Germans 2.0% Chuvash 0.6% Others 4.5%
Republic of Tatarstan	3,773,800	70	Tatar 48.5%	43.3%	Chuvash 3.7% Ukrainians n/a Udmurts n/a Mordva n/a Mari n/a Bashkir n/a
Sakha Yakutiya	1,094,065	80	Yakut 33.4%	50.3%	Ukrainian 7.05% Tatar 1.6% Belarusian 0.9% Buryat 0.77% Bashkir 0.38% German 0.37% Moldavian 0.34% Chuvash 0.31%

Table 2.1 Continued

Region	Population	Ethnic composition			
		Number of nationalities	*Percentage of indigenous population*	*Percentage of Russian population*	*Percentage of other major nationalities*
Astrakhan Oblast	1,032,800	150	Russian and other Volga Muslim nationalities n/a	72%	Kazakh 12% Tatar 7%
Republic Ingushetiya	308,000	60	Ingush 83.7%	14.3%	Chechen 10.6% Other 1.7%
Republic Mai El	758,900	Over 50	Mari 43.3%	47.5%	Tatar 5.9%
Republic of Komi	1,200,000	100	Komi 26%	62%	Ukrainians n/a
Republic of Tyva	311,200	50	Tyvin 67.1%	30.4%	Others 2.5%
Udmurtiya	1,644,000	n/a	Udmurt 31.9%	59%	Tatars 7%

improve the republic's relations with the Kremlin and enhance the republic's position vis-à-vis the centre.[20] Similarly, political connections between the former Minister of Transportation Nikolai Akseneko and regional governors often led to the implementation of important economic projects in chosen regions.

Second, despite regional dependence on the centre in some key areas, the Yeltsin era was generally characterised by the political weakness of the presidency and internal struggles between various financial groupings within the Kremlin. These factors precluded the centre from formulating a durable coherent policy towards the regions. This logically resulted in a gradual increase in the influence of regional elites at both national and regional levels throughout the years 1997–9.

Remarkably, the political weakness of the centre and its inability to contain growing political and economic asymmetry between the regions evoked a generally negative reaction in the provinces, fomenting complaints by regional experts and politicians. The latter began to revise the country's federal relations towards a more coherent and stable framework. In short, because the centre could not deliver a sound regional policy, the federation's subjects slowly began to take the initiative into their own hands. Regional political elites emphasised that the existing framework of centre-regional relations was no longer sustainable and that both federal and regional governments should change their mode of interaction. Despite the fact that the regional leaders did not compose a unified programme for reforming the system of centre-periphery relations, the range of ideas that were discussed demonstrated the willingness to pursue change.

The governor of Saratov Oblast Dmitrii Ayatskov, for example, declared that instead of developing a treaty-constitutional federal system, the national government should embark on establishing a single 'power vertical, in which governors would be appointed officials, accountable to the president and the government'.[21] Egor Stroev, governor of Oryol Oblast, adopted a more moderate line, suggesting that the centre must introduce a range of laws providing for modification of regional legislation if the latter were found to be in contradiction of the federal constitution.[22] The Mayor of Moscow, Luzhkov, indicated that the unregulated nature of centre-regional relations in conditions of permanent economic crisis increased the tendency toward political and economic self-isolation on the part of the regions. Luzhkov suggested that Russian centre-regional relations must be based on unified socio-economic and institutional principles and above all on the system of unified fiscal policy.[23]

Similarly, the governor of Samara Oblast Konstantin Titov presented his plans for restructuring centre-periphery relations, underscoring the importance of formulating effective federal legislation, and institutionalising the division of responsibilities and properties between the centre and the regions.[24] The governor of Primorsk Krai Evgenii Nazdratenko declared on a more radical note that by 1998 'potential disintegration of the Russian Federation became a real threat as the regional governments are one by one boycotting federal taxes'. Nazdratenko placed the blame for this 'catastrophic situation' on the federal government, which 'systematically created grounds for the country's collapse by promoting the differing economic and political rights of the regions'.[25] The governors of many other influential regions such as Krasnoyarsk Krai and Orenburg and Sverdlovsk Oblasts echoed many of these ideas in principle.

Furthermore, academics and politicians in deprived recipient regions (Kozlov 1998, pp. 49–57) were alarmed about the fact that the centre provides no incentives for centralisation and even suggested that in the economic and financial spheres, the Kremlin purposely pursued a policy of promoting centrifugal and separatist tendencies. In particular, they looked at economic problems as a starting point for the future revision of current federal policies. Republican leaders and experts (Khakimov 1998, p. 45) also favoured the change: not only did they reiterate that their respective republics wished to remain within the federation but they also argued that 'there is a need to begin reciprocal movement by the centre and subjects of the Federation towards creating a single legal space'. Even Shaimiev, President of Tatarstan and the most independent Russian regional leader, later insisted that 'Putin's initiatives were strongly expected; for me personally, nothing unexpected happened. I knew that as soon as the new president comes, he would simply be obliged to strengthen federal power, as he should advance reforms and achieve well-being in Russia'.[26]

Statistical data further confirms the existence of a consensus among the regional leaders on the necessity of reforming their relations with the federal centre. Opinion polls conducted at the Centre of Sociology of Ethnic and

Regional Relations in 1999 mirror the statements given by politicians. Re (see Ivanov and Yarovoi 2000, pp. 43–54) demonstrates that a majority of regional elites (50–60 per cent) negatively evaluated the weakness of the federal centre's policy towards their respective regions. Moreover, a majority of regional experts negatively evaluated the existing institutional forms of regional relations, and in particular the centre's initiative in concluding bilateral treaties. According to the polls, 22–35 per cent of the experts insisted that the treaties represented a threat to Russia's territorial integrity; 45–60 per cent suggested that the treaties were a necessary measure at the time but should cease to exist in the future; only 12–25 per cent thought that bilateral treaties represent a modern form of federal relations (Ivanov and Yarovoi 2000).

This overwhelming consensus on the need to restructure the framework of centre-periphery relations led to a parliamentary agreement in June 1999 to introduce the law 'On the Principles and Order of the Division of Responsibilities between the Federal and Regional Branches of Power of the Russian Federation'. This document demanded that the regions comply with the principles of supremacy of the federal law and constitution, and ensure the equality of all subjects of the federation in their relations with the centre (Vishnyakov 2002).

Third, the policy weakness of the Kremlin vis-à-vis the regions was particularly visible when the federal centre formed a coalition government in September 1998 headed by Primakov, who proceeded to consolidate regional politicians and business elites – in particular *Lukoil* and its dependent companies – into a powerful political movement *Otechestvo-Vsya Rossiya*. Apart from the Prime Minister, the leaders of the movement were Mayor of Moscow Luzhkov, President of Tatarstan Shaimiev, and Mayor of St. Petersburg Yakovlev. OVR advocated centralisation of federal power through restricting regional independence. Primakov, for example, insisted that there was a need to preserve the country's territorial integrity, put an end to moves towards separatism by republican leaders, and work towards the creation of a single legal space across the country through the revision of legislation negotiated in the existing bilateral treaties. To Primakov's mind, the Ministry of Justice was supposed to observe legislation adopted in the regions and monitor its compliance with the single federal standard.

The Prime Minister also advocated greater universalism in centre-periphery relations and criticised the cliental tactics of his predecessor Viktor Chernomyrdin. At its first Congress held on 24 April 1999 in Yaroslavl, OVR called for a restoration of the *power vertical*, which was manifest in the dual responsibility of the regions and the centre before the public.[27] A so-called *strategic equilibrium* system was part of the newly proposed *power vertical*, which meant that the delegation of authority to the regions could only take place if the regions assumed greater financial and socio-political responsibility.[28] Undoubtedly, OVR had ambitions to seize power and ultimately take the lead in implementing this programme.

The emergence of OVR and its subsequent political competition to the Kremlin created a split within the central elite on the issue of handling regional

politics and regional parties. For example, Dyachenko, Yumashev and Berezovskii regarded Primakov's political coalition as a direct threat to presidential
authority and purposely supported a rival to the OVR regional formation *Golos
Rossii* established by the ex-governor of Samara Oblast Konstantin Titov (Guseinov 1999, p. 421). On the other hand, Chubais's grouping, which at the time
included Putin and was backed by the *Lukoil* lobby, advocated a more 'moderate' approach and was prepared to support OVR with certain reservations.[29] It is
due to these developments that the Deputy Head of the presidential administration Sysuev – one of the main representatives of the Chubais clan – announced
that Primakov's government represented a strategic ally of the Kremlin and
resigned following Primakov's dismissal.[30]

The OVR issue underscored the federal centre's weakness precisely because
of the fact that the regional policy advocated by the OVR and the Kremlin's own
aspirations towards reforming existing centre-periphery relations ultimately converged. That the centre was unable to pursue its policies in the regional dimension and permitted a rival to capitalise on its ideas demonstrated the lack of
political consolidation within the Kremlin and its weakened authority. The following examples of the Kremlin's proposed policies demonstrate the ultimate
theoretical similarity between the views of the centre and OVR on reforming the
system of centre-regional relations.

As early as 1996, Yeltsin's analysts were elaborating on a possible and pain-
less transfer from the existing scheme 'one region – one presidential representative' to a more manageable system of enlarged districts, in which presidential
representatives would be disentangled from the influence of the governors. The
initial offers on the number of such districts varied from 24 (suggested by
Chubais) up to 34 (suggested by Smirnyagin, Satarov, and Pain) (see Lysenko
2002, p. 40). During 1997, Chubais's administration continued to draft measures
to increase the role of the presidential representatives in the regions. The government even adopted a new bill aimed at enhancing the performance of this institution. However, the weakness of the centre and the absence of regional cadres
capable of counterbalancing the influence of the governors precluded the implementation of this plan. The centre simply could not find sufficient personnel in
the regions who would be able to act independently of the incumbent governors
(Lysenko 2002; Guseinov 1999, pp. 303–4).

The situation regarding Primorsk Krai demonstrated the centre's failure to
implement a programme to strengthen the institution of presidential representatives. The appointment of the regional head of the FSB General-Lieutenant
Viktor Kondratov to the post of presidential representative in May 1997 evoked
a number of far-reaching conflicts. These disputes have spilt over to the wider
public arena and the judiciary and acquired a national dimension when the State
Duma alleged that the presidential representative overtly supported the pro-
Yeltsin Mayor of Vladivostok Viktor Cherepkov in his political conflict with
Evgenii Nazdratenko.[31]

Despite the apparent failure in Primorsk Krai, the centre was determined to
increase the number of pro-Kremlin politicians in the provinces. The strategy of

incorporating opposition governors elected in 1996–7 into the existing power elite confirmed the Kremlin's willingness to make the centre-periphery system more manageable and to take firmer control over the governors. Given that many regional leaders and their teams came to power through anti-governmental or Communist electoral campaigns and following these elections centre-periphery relations were at their lowest ebb, the centre rapidly changed tactics. Instead of exacerbating the situation, it took a step forward and embarked on disentangling the newly elected governors from the opposition, taming them, and turning them into organic members of the 'party of power'. Guseinov (1999, pp. 298–304) extensively describes this dynamic. He notes that the situation in Kemerovo – a coalmining province – proved this point. Yeltsin fully supported the election of his number one opponent Aman Tuleev in July 1997 and in October dismissed his own representative in the region Anatolii Malykhin, Tuleev's long-term adversary. Similarly, in January 1998, the governor of Leningrad Oblast Vadim Gustov – an outspoken opponent of Yeltsin – received support from the centre to implement strategic port-building projects as well as an Award of Honour as a symbolic gesture of the Kremlin's goodwill. In September 1999 the then Prime Minister Putin paid a special visit to opposition governor Leonid Gorbenko in Kaliningrad.

Furthermore, in order to avoid a repetition of the political defeats of 1996–7, in October 1999 Yeltsin established a co-ordinating committee *For Fair Elections*, the main function of which was to sabotage the electoral campaigns of governors representing rival parties and movements. Georgii Satarov, a former advisor to Yeltsin, went to head the committee whilst other important actors in the institution represented the Union of Russia's Journalists – a mass media organisation dominated by persons close to Kremlin circles. The committee sent a large number of its representatives to the 'red belt' regions (the areas of central Russia dominated by the Communist Party) and employed some dubious strategies, such as launching a competition for the 'fairest' and 'dirtiest' electoral campaigns.[32]

All these moves served to demonstrate the centre's determination to become more closely involved in regional politics and to establish – even through an administrative mechanism – a common ground among federal and provincial politicians. These developments underscored a convergence of the Kremlin's ideas with those of the OVR – in particular on the point of creating an executive 'power vertical' and consolidation of the centre's power over regional politics. These developments demonstrate that, by the end of the Yeltsin presidency, the normative vector of Russia's federal relations had unequivocally pointed in the centralising direction. This represented a marked difference from the early period of unlimited sovereignisation and chaos.

The dismissal of Primakov's government in May 1999 and the Kremlin's choice to rival the OVR ensured that the ideas of neither the Kremlin nor Primakov-Luzhkov-Shaimiev on reforming the federation were actually implemented until Yeltsin's successor decided to incorporate them into his programme of federal restructuring. In this context, it is important to remember that, because

Putin's official duties during his service in Yeltsin's administration were related to the sphere of centre-regional relations, he had the institutional ability to capitalise on existing proposals for reforming the federal system. Indeed, on Putin's arrival in the presidential administration in June 1996, he occupied the post of member of the administration's territorial department responsible for reforming centre-regional relations. Putin continued to manage centre-periphery affairs in his later post as deputy head of the presidential administration (from March 1997). Following his appointment as chief of the FSB on 25 July 1998, Putin maintained close control over centre-periphery relations. At the time, he organised a special department for centre-regional co-operation and launched a number of investigations to combat corruption in Russia's provinces. Undoubtedly, the massive amount of information that Putin managed to accumulate and had to hand during his three years of tackling the problem allowed him to shape his federal programme from unimplemented proposals produced by both the OVR and the late Yeltsin administrations.

Socio-psychological and cultural factors

Apart from the socio-political factors that underpinned Putin's centralising efforts, there were a number of socio-psychological, cultural causes. Putin's re-centralisation has been sustained by the reluctance of political society to opt for liberal democratic solutions and by public expectations at the end of the 1990s. In this case, the forces of Russia's political style and culture have come to the fore.

First, the lack of appreciation of democratic and federal principles on the part of the federal elite led to the emergence of various misinterpretations of democratic ideas and finally to support for re-centralisation as the only viable and familiar solution. In this sense, when faced with the need for reform it was easier for many to revert to semi-authoritarian centralising policies than to attempt to comply with democratic models borrowed from the West. A large number of Russian political leaders and their advisors were unable to regard democracy as 'the only game in town' simply because they were not born within a democratic state and they had to re-adjust their norms, values, and beliefs in the middle of their adult lives. Colton and McFaul (2002, pp. 117–18) go a step further by suggesting that Russia's difficulties in consolidating democracy lie within the 'inadequately democratised institutions', whose leaders behave more like 'elected tsars' and even lag behind the population in terms of assimilating democratic values.

These stylistic constraints and patterns make no surprise of the fact that consolidation of the federal government's authority evoked positive responses among many political analysts close to the Kremlin. Putin's former cabinet advisor Vadim Pechenev (2001, pp. 41–2) wrote that 'a strong state must be created in Russia before everything else – federation, democracy, etc. Whether others like it or not, these are the fundamental basics of this one-sixth of the planet'. Other analysts such as Alexandr Solzhenitsyn, Andranik Migranyan,

Mikhail Leontev, and Vyacheslav Nikonov echoed these thoughts (though with differing interpretations), primarily insisting that Russia's only real option was to go through a phase of 'progressive' (i.e. economically viable and socially stable) authoritarianism before being able to build a genuinely democratic system.[33]

More importantly, the composition of the regional political elite has not changed in principle since the Soviet era. Researchers at the Institute of International Studies (Nichoson 1999, p. 27) have noted that, while ministers of the former Soviet Union, with the exception of Chernomyrdin, ceased to occupy key positions at the centre, most of the 'old guard' retained their provincial posts and old managers continued to determine the nature of regional and local affairs. Lane and Ross (1999) reached a similar conclusion, pointing out that key members of the Soviet economic and administrative elite began to dominate Russia's local assemblies. Badovskii and Shutov (1995) also insisted that institutionalisation of post-Soviet regional elites represented the process of transformation of the old *nomenklatura* into the 'party of power'. This situation ensured that regional political elites were psychologically prepared for a return to centralising policies as the most natural and traditional response to existing socioeconomic problems.

The leadership of the old elites was particularly prominent in the three most important areas: gubernatorial elections, legislative elections, and appointment policy in the executive office. It is clear that gubernatorial elections during 1996–7 brought no democratic change in the composition of the elites. As mentioned earlier, campaigns took place under the banner of opposition to the federal centre. However, this did not mean that those favouring genuinely democratic reforms seized key positions in the regions. Russian experts have estimated that two thirds of the new governors who came to power with the support of the opposition, were members of the CPRF (Communist Party of the Russian Federation) and the NPSR (Popular Patriotic Union of Russia). The conservative opposition was able to celebrate when the Communist Yurii Lodkin became the governor of Bryansk Oblast; Tuleev won the governorship in Kemerovo Oblast, Gustov in Leningrad Oblast, and many others – 'red belt' governors – were elected in the Central-Black-Earth Regions of Russia (Guseinov 1999, pp. 294–6). Following the elections to regional legislatures that took place mainly during 1998, the Communist opposition further confirmed its position as the single influential party in Russia's regions. In 1998, the CPRF-NPSR won 40 per cent and 18 per cent of deputy seats in Stavropol and Khabarovsk Krais respectively; 18 per cent in Astrakhan Oblast; 37 per cent in Belgorod Oblast; 12 per cent in Moskva Oblast; 35 per cent in Novosibirsk Oblast; 38 per cent in Penza Oblast; 33 per cent in Smolensk Oblast, and 28 per cent in Tambov Oblast (Petrov and Titkov 1999).

Besides the electoral successes of the non-democratic opposition, the appointment policy in executive offices of many regions effectively precluded an influx of new non-*nomenklatura* cadres to regional politics. The socio-political situation in the majority of Russia's regions was marked by social apathy, the over-ambitious ethnic policy of regional leaders, pseudo-independent institutional structures and the weakness of the federal centre. This created grounds for the 'crystallisation' of the current composition of the ruling political elites, which in turn led to the emergence of strong patron-client links and the system of power 'inheritance' (Kuzmin 2001). Two regions – the Republic of Tatarstan and Penza Oblast – serve as excellent illustrations of the appointment policy in executive offices in Russia's regions. According to research carried out by Kazan State University's social studies department (Kazakov 2001), in 1996–7 the political elite of the Republic of Tatarstan included 84.3 per cent of representatives of the former party *nomenklatura*. The President of the Republic, Shaimiev, had formerly occupied the post of First Secretary of Tatar Oblast Executive Committee of the CPSU (*obkom*); the head of the presidential administration had headed the party administrative department on the same Oblast Executive Committee of the CPSU. Similarly, the speaker of the republican parliament had formerly headed an administrative department of the Tatarstan Oblast CPSU Committee and the Prime Minister was a former head of the city committee of the CPSU in the republic. Penza Oblast largely repeats the picture. The estimated share of the former Soviet *nomenklatura* in 1996–8 in the regional administration composed 96–7 per cent, while new personnel unrelated to the former Communist leadership occupied non-influential positions playing marginal roles in regional developments. Regional institutions of the executive power were staffed with former heads of *raion* party administrations, and heads of various CPSU branch departments and the like. Mendras (1999, pp. 304–6) has also pointed out that a majority of Russia's regional elites not only comprised 'old guard' personnel, but that this elite also carefully shielded itself from any outside interference by those not belonging to the current establishment and the existing networks. Analysis of elite behaviour in Ulyanovsk, Bryansk, Smolensk, Magadan, Orenburg, Omsk and Sverdlovsk Oblasts proves this point.

In addition, the fact that the scientific intelligentsia, largely without special knowledge of political science, often occupied influential positions in local government and local representative organs, diminished the prospect of a deeper theoretical understanding of democratic principles. The professional orientation of members of local Soviets in Moskva Oblast, for example, lied outside politics or even the social sciences: engineers, natural science university teachers, doctors, and agricultural managers dominated lists. There were virtually no representatives of the legal profession and very few of these politicians were party affiliated.[34] Lallemand (1999, p. 314) has reached a similar conclusion in his analysis of the professional background of elites from the regional administrations of Bryansk and Smolensk Oblasts. He insists that 'an emphasis on the managerial pragmatism of the administrative elites – which had its deep roots in the Soviet system – naturally resulted in the search for cadres with a non-sociological educational background'.

The pattern of poor political knowledge and the traditional elite's propensity towards authoritarian political style fits perfectly into the summary of regional political elites' mentality produced at the Institute of Sociology (RAN) by

Chirikova and Lapina (1999, pp. 116–23). The authors demonstrate that regional business and political elites have not developed stable political orientations, and only a limited number (one-fifth) of elites were party affiliated. The sociologists drafted five differing models of elites' political psychology. Four out of these five models represented either a passive or an authoritarian approach to politics. These behavioural models included total political non-engagement, orientation toward a strong national leader (Luzhkov, Zyuganov etc.), political dependence on the highest echelons of regional elites, and political adaptation manifest in the alignment of views with the current elite. Only the final model, *political involvement*, highlights the attempts of regional leaders to create parties of power in order to secure authority and the likelihood of victory in forthcoming elections.

Second, public support for re-centralisation represents another fundamental aspect of the socio-political factors under discussion. The years of Yeltsin's 'liberal-democratic' rule ensured that the population was prepared to opt for socio-political and economic stability at the expense of abstract democratic notions and their unlikely future gains. Mere disillusionment with what many reformers called a 'liberal-democratic transformation' only enhanced traditional Russian affection for strong central power capable of enforcing order. Colton and McFaul (2002, p. 91), in suggesting that a dichotomy between democracy and order need not exist, nevertheless quote important VTsIOM polls conducted in January 2000. The data shows that 75 per cent of Russians were 'in accord with a statement that order is more important than democracy and should be pursued even if it entails violations of democratic procedures and abridgement of personal freedom'. Furthermore, research polls conduced by the Centre for National and Regional Relations (RAN) (see Ivanov and Yarovoi 2000, pp. 123–62) indicate that in 1999 the population was expecting consolidation of central power as a prerequisite for extinguishing socio-political turmoil. Nearly 53 per cent of respondents agreed on the necessity of strengthening Russian statehood through reform of the existing administrative mechanisms of centre-periphery relations. 40 per cent insisted that reforms must include the elimination of subjects' differing constitutional status; 41 per cent envisaged changes in the acting constitution to achieve this goal; and 53 per cent suggested that creating a strong 'power vertical' would help to consolidate the integrity of the Russian Federation. It is therefore not surprising that VTsIOM polls conducted on 26–29 May 2000 registered a 48 per cent approval rating for Putin's proposed federal reforms and 68 per cent approval for the consolidation of central power associated with it.[35]

The discussion above demonstrates that the country was prepared socially, politically, and institutionally to enter a period of re-centralisation. Summarising the political mood in the regions and at the centre in 1998–9, Chirikova and Lapina (1999, pp. 182–6) suggest that the biggest social paradox of the time emanated from the fact that the regions needed the centre and the centre needed the regions. However, the uncertainty of the centre's strategy and the political immaturity of regional elites made it difficult to determine the scope and sphere of co-operation between the two poles of power. In brief, at the end of the 1990s, the regions waited for new central policies but the centre could not yet deliver them. With the election of a new president in March 2000, this situation changed.

Putin's early tactics and strategies

Despite the socio-political and ideological predisposition of political society towards enactment of the federal reform, it could not have been launched without some form of political struggle. Therefore, it would be a simplification to insist that the centralising turn in its present shape and form unequivocally represented a logical move predicted and desired by many. Regardless of the fact that many of Putin's federal ideas emerged during Yeltsin's era and socio-political conditions were conducive to centralising trends, regional leaders did not practically expect the changes to be so radical. Indeed, during debates throughout the period 1998–9 many politicians, while agreeing on the necessity of revitalising federal power, were far from assuming any radical revision of the regions' existing legislation in favour of federal law and no less radical reform of the Federation Council. A number of governors emphasised the inefficiency of other regions' regulations, suggesting the need to raise the rights of remaining territories to the level of the existing republics. They also accentuated the need to enhance manageability of institutions not at their own level but at a level above or below their respective spheres of influence, i.e. governors mainly pointed to local mayors as a source of problems whilst mayors pointed to the governors (Chirikova and Lapina 2001, p. 24). These views were explicit in the ideas of the governor of Sverdlovsk Oblast Eduard Rossel, who insisted on dividing Russia into purely territorial republics based on a strong 'power vertical' at the local level only.[36] Alexandr Lebed, the governor of Krasnoyarsk Krai, while seeing Russia's biggest problem in unravelling the federal 'power vertical', saw the solution in limiting the constitutional role of the national government to common federal duties such as military and foreign policy and leaving other initiatives entirely to the regions.[37] Farid Saphiullin, the third Duma MP elected from Tatarstan, also insisted that 'while the necessity of federal administrative reforms is clear to all, there is still an alarming possibility that such reform could go too far without the consent of the regions, the public, and without compliance with already established federal tradition'.[38]

The political and institutional arsenal with which regional leaders could have attempted to overturn Putin's projects was rather impressive. Institutionally, membership of the Federation Council guaranteed gubernatorial immunity and according to the constitution, the senators had the right to block bills that threatened the existing balance of centre-periphery relations. In addition, the political situation after the president announced his plans for administrative and federal reforms became extremely volatile. The elite coalition that elevated Putin to power realised that he refused to play the role of a marionette simply representing the interests of the 'family'[39] and maintaining the political stage of the Yeltsin era. It was then clear to everyone that these forces were trying to put the brakes on a president who had made his first moves independently of Yeltsin's former clan. Therefore, during May-August 2000 the governors could have over-turned Putin's initiatives by relying on the emergence of a credible opposition to Putin, which officially resisted the consolidation of a super-presidential regime combined with overwhelming importance of federal power. The opposition to Putin comprised four major groupings: (1) 'the family', which was not interested in a strong president and wanted to maintain a fragmented political stage, (2) representatives of big business, who were alarmed by the prospect of a large-scale redistribution of property, (3) the governors, who were unwilling to give up their powers, and (4) the liberal SPS (Union of Right Forces) and *Yabloko* parties, which were financed by various economic groupings.

Boris Berezovskii, Russia's then most powerful business executive and the former head of the Security Council, became the most outspoken representative of the opposition. Berezovskii's letter, written just a day before the Duma voted on Putin's federal proposals, demonstrated that the tycoon had declared open political warfare on the presidential administration and assumed the role of an advocate of gubernatorial interests in the State Duma. In his statement, the oligarch overtly criticised the president and insisted that his proposed administrative and federal reforms were 'directed toward changing the state's structure' and represented a 'threat to Russia's territorial integrity and democracy'.[40] Berezovskii also announced his own proposals to 'strengthen the federation and power', including abandoning proposals to create seven federal districts, instituting direct elections to the Federation Council and boosting regional independence within a uniform federal legislation. The tycoon ultimately wanted these issues to be put to the vote in a referendum, which would clearly serve the interest of Putin's opponents who owned the most important media channels and could therefore influence public opinion prior to the ballot.

Despite this, by the end of July 2000 Putin had managed to defeat the efforts of the opposition and consolidate his authority. With this victory, Putin established himself as an effective political tactician 'capable of resolving complex problems efficiently on several parallel fronts' (Drozdov and Fartyshev 2000, p. 239). His simultaneous actions ensured political victory over those who opposed or would potentially oppose the president's plans for a new federal and administrative design. No less important was the fact that he secured Western approval to enforce the administrative drive against financial crime – fraud, tax evasion, and money laundering – which were often associated with 'family' politics.[41] Other explanations of Putin's success include a number of apparent but nonetheless important aspects: a high degree of popular trust of the centre, a lack of co-ordinated regional policies, and uncertainty of future gains in the case of adopting active opposition to the reforms, weakness of political society, and the fear of coercion. We can suggest five major aspects that shifted the political balance towards Putin's advantage: (1) appointment policy; (2) international recognition; (3) the arrest of Gusinskii; (4) institutional compromises between Putin and certain members of the opposition; (5) the success of federal forces in Chechnya.

First, in his immediate choice of appointments the president demonstrated that he wished to rely on people with whom he had co-operated in the past and who would be loyal to him. Despite some appointment restrictions informally imposed by Yeltsin in his first presidential year, Putin nominated his most trusted people to key positions in the military, the Security Council, and the FSB immediately after taking office. The head of the FSB Nikolai Patrushev was a colleague of Putin's in St. Petersburg, and had worked for the KGB combating corruption in the city during 1972–92. Putin chose Sergei Ivanov to head Russia's Security Council before his transfer to the MOD in March 2001. Having studied together at Leningrad State University, Putin and Ivanov became close friends whilst serving as members of the KGB in the city. Ivanov openly declared himself a member of Putin's team and demonstrated his sheer indifference to the opposition. In an interview with *Komsomolskaya Pravda*, he insisted that he had not read Berezovskii's letter to Putin because 'it is physically impossible to read everything in the printed media', purposely disregarding the fact that the letter reached the front pages of all the national newspapers.[42] Furthermore, despite the failure to replace Minister of Defence Igor Sergeev, the presidential representatives to the regions were mostly high-ranking military officers – five out of the seven were generals. In order to intertwine his appointments institutionally, Putin made plenipotentiary representatives members of the Security Council. Besides effective appointments in the government, Putin managed to attract the support of the country's richest company and the main taxpayer *Gazprom*. Prior to the arrival of Alexei Miller as its chief executive in April 2001, Putin appointed a former St. Petersburg colleague Dmitrii Medvedev to head the board of company directors in June 2000. Apart from being the major revenue provider, *Gazprom* was also an important political actor. The state employed *Gazprom's* financial hold over *Media-Most* to step up its political pressure on the opposition, whose other main member was Vladimir Gusinskii. *Gazprom*, which already owned a 14 per cent stake in *Media-Most*, demanded the return of a further 20 per cent of the company as collateral for a $211 million loan that expired in March 2000, which virtually destroyed Gusinskii's financial empire and deprived the opposition of its voice on the state's main channel.[43]

Second, given that Putin has chosen to rely on forces that could launch attacks on the opposition on legal grounds, he had to explain this policy to the international community. The legal enforcement agencies, which stood behind the president, targeted Russia's industrial elites and attempted to place their attacks into the framework of the international campaign against money laundering and the misuse of IMF lending facilities. By doing so, they attempted to compromise these oppositional figures in the eyes of the West and make them scapegoats for the mismanagement of IMF money and state funds during the past decade. The international climate in the summer of 2000 was supportive of such a task. All major international meetings during the summer involving Russia, including the visit by President Clinton to Moscow on 4–5 June 2000 and the G8 summit in Okinawa on 21–23 July 2000 (and even the earlier visit of NATO General Secretary George Robertson to Moscow on 15–16 February 2000), discussed money laundering as one of the most serious problems threatening international security. Because Russia's presence at the Okinawa summit would have made discussion of such issues very sensitive, US President Bill Clinton visited Moscow a month before the beginning of the G8 Summit. During this meeting, the American and Russian leaders spent much of their time discussing economics. Employing the international momentum to assist his drive against the oligarchs, Putin reassured the US president that Russia would take a tougher stance towards money laundering and presented an efficient economic recovery plan, which included strong measures to protect property rights, to ensure tax returns, and to combat bureaucratic corruption.[44] Clinton's positive evaluation of Putin's attempts to reform the state and economy signalled the West's 'green light' for the new Russian government and its administrative policy. The G8 Summit in Okinawa represented a final accord in the creation of a more favourable international climate for Putin's domestic policies: the G7 released a report in which the participant nations criticised Moscow for a lack of adequate measures to combat the processing of illegally acquired money.[45] All these initiatives gave full political backing to Putin's campaign against the opposition at home.

Third, the arrest of Media-Most owner Gusinskii on 14 June 2000 (ten days after Clinton's visit to Moscow) on suspicion of having played a role in the theft of $10 million of state money in a privatisation deal was an organic part of the campaign against money laundering, corruption and fraud. Despite the fact that this arrest attracted a lot of attention abroad and among Russian liberals, this event failed to foment popular protest and to raise the profile of Putin's opponents. Interestingly, following his arrest Gusinskii's own rating fell, while Putin's remained at the same level: 10 per cent of the respondents to a poll conducted by the *Indem* fund began to view Putin's policy more favourably; 12 per cent worse; and 58 per cent had not changed their attitude to the president at all.[46] The head of VTsIOM Professor Yurii Levada (2000) also observes that for the majority of the population Gusinskii became an 'anti-hero': 25 per cent of Muscovites were satisfied with the tycoon's arrest and 87 per cent genuinely believed that he was involved in 'some kind of fraud'. Russian political circles also failed to employ the Gusinskii affair in order to strengthen the standing of the opposition. In particular, Russia's liberal right forces demonstrated their political weakness and inability to form a united opposition. Immediately following the event Grigorii Yavlinskii, the leader of *Yabloko* party, became an outspoken critic of the president. Yavlinskii insisted that the arrest of the oligarch was on the direct orders of the Kremlin and was aimed at curbing the freedom of the press. He called for the creation of a united anti-fascist opposition from '*Yabloko* to the CPRF'.[47] The result was the announcement of a provisional merger between *Yabloko* and SPS. However, against a backdrop of clear public indifference, even a positive evaluation of the Gusinskii affair, Yavlinskii as well as the SPS rushed to disentangle themselves from the open opposition. Instead of the promised merger, on 21 June 2000 the two parties signed a dubious non-binding document that aimed to create a united democratic coalition of right forces within one or two years.[48] It is also indicative that Yavlinskii himself as a leader of *Yabloko* did not put his signature to this document and delegated the responsibility to a high-ranking party official, Vladimir Lukin. This clearly meant that *Yabloko* – despite the fact that it was financed by Gusinskii – decided not to go too far in its opposition, particularly as to do so could have damaged its rather weak rating.

It has now become clear that, apart from the main actors, Berezovskii and Gusinskii, other opponents of Putin – the Duma and its member parties the CPRF, SPS, *Yabloko*, the governors, and some oligarchs – were prepared to compromise with the president on various matters. This was particularly true if these moves implied a long-term political gain. Putin was also open to such compromise. Given that he had to neutralise the mounting opposition financed by serious political rivals, he proceeded to fight for all available political actors by offering them certain advantages in exchange for support for his policies. First, he played on the State Duma's institutional ambitions as well as on its traditional reputation as the centre of anti-Yeltsin opposition. It is within this context that Putin's displeasure with former Yeltsin supporters and the system of oligarchic capitalism evoked positive responses among anti-Yeltsinists in the lower house of the Russian legislature. Moreover, the State Duma as an institution had strong incentives to support Putin's initiatives in order to improve its position vis-à-vis the Federation Council. A positive vote on presidential proposals would enable the lower house to weaken the authority of its colleagues in the Federation Council and create a new balance of power, in which an emboldened Duma and the president would oppose a weakened Federation Council and the governors.

Indeed, in the conflict between the two houses of parliament over the presidential proposals on the reform of the regional system both the Duma and the president demonstrated an insulting attitude towards the Federation Council. Presidential representative in the Federal Assembly Alexandr Kotenkov declared that 'having lost deputy immunity a large number of the governors will end up in prison the very next day'.[49] The leader of the LDPR Zhirinovskii supported his words by mainly attacking the 'red governors', while the leader of the Union of Right Forces Nemtsov insisted that the deputies would override the Council's vetoes by all means necessary – 'regardless of the opinion of the upper house'.[50] Moreover, some Duma deputies suggested that the new regional representatives – due to their inherently lower status as appointed second ranking officials – should no longer be entitled to decide on such serious matters as Russian foreign policy and the declaration of a state of emergency.[51] A number of regional representatives working with the State Duma on the matter complained about an 'undisguised hostility towards the Federation Council'.[52] It was partly thanks to these developments that the three bills introduced by President Putin aimed at strengthening federal government control over the regions received overwhelming approval during the first reading with 360 positive votes.[53]

It now becomes clear that Putin's strategic alliance with the Lower House of parliament played an important role in his struggle for influence and ultimately in the enactment of the package of federal reforms. Finally, the failure to capitalise on the existence of credible opposition to Putin led the governors to support regional reform in principle. In order to protect themselves from the consequences of possible coercion and political embarrassment the governors preferred to vote for all the proposed reform principles including the new method of formation of the Federation Council. As Putin carefully put it in official language, the gubernatorial decision to back the reform package stemmed from their political 'responsibility' and the desire not to lose the reputation of a historical stabilising force in Russian politics, which 'prevented the country from political and territorial disintegration during the turbulent 1990s'.[54] In Putin's words, the governors agreed to reform the federation because 'they understood that unregulated relations in the federal sphere impeded the development of not only the entire state but also their respective regions. This was a responsible decision and this decision was theirs'.[55]

Finally, the success of federal forces in Chechnya represented a major contribution to the consolidation of Putin's authority and to his increasing popularity. This has also contributed to the poor public evaluation of the old pro-Yeltsin elite, which was supposedly to blame for the outbreak of the first Russian-Chechen war and the subsequent failings of the Russian military.

3 The centre and the regions

A new balance of power

The centralising direction of the federal policy vector that had taken place by the end of the Yeltsin era did not automatically guarantee that President Putin and his team were able to implement federal reforms in their original form. Various external factors such as the resistance of involved actors, the scarcity of their power resources, and inertia of the previous regime could frustrate new policies. From this point of view, the interaction between the structure and the process of Russia's federalism seems particularly interesting.

The centre attempted to secure its grip over the regions by introducing a range of substantial institutional alterations aimed at centralisation and hierarchy. The regions, on the other hand, responded by adapting these measures to local realities in an attempt to preserve the main dimensions of federal rights and freedoms that had been established before the enactment of the reforms. Thus, the apparent rigidity of the ongoing structural changes failed to eliminate the pre-existing centre-regional bargaining and local aspirations for a greater political self-determination. As a result, the pre-existing fabric of federal relations did not disappear. Rather, it became disguised under the rigid structural façade permanently exerting substantial pressures towards creation of a new balance.

One indication of this dynamic is that the newly established centre-regional balance of power followed a combination between the 'elite settlement' and 'winner takes all' scenarios discussed by Gelman (2000, pp. 236–8) in his comparative analysis of Russia's regional reforms. The 'elite settlement' strategy takes place 'if the dominant actor has insufficient resources or other limitations to the use of force, while his competitors have enough resources for survival, but not enough to become the dominant actor. Thus, both sides benefit from an "elite settlement": the dominant actor secures his position, while his competitors receive access to subordinate positions within the governing group'. In the 'winner takes all' scenario, the previous regime does not disintegrate fully, formal institutions operate mainly as a façade, and the dominant actor does not have any obstacles to excluding other actors from the political process and securing direct or indirect control over political society and the media. Russia's regional transformation indeed exhibited features of both these scenarios.

First, analysis of central and regional power resources shows that at the beginning of the reforms, both levels of government had solid levers of influence and none of the governmental structures, the centre or the regions, was politically capable of completely overshadowing its opponents – hence the necessity of an 'elite settlement'. Second, further evaluation of the reforms' outcomes suggests that the initiated programme had a limited scope. Participants in the political process reached a reasonable compromise in a number of important policy areas. Beside the compromise, many subjects were able to retain their legal-political positions and previous power balances perpetuated, masked under the guise of new verbal formulations. Such an elite settlement illuminated the emergence of adaptive, non-transparent forms of Russia's federal processes. At the same time, in conceding to provide certain privileges for certain regions the centre placed itself at the top of the new 'redistribution tree' and enhanced its control over the socio-economic activity of the regions thus turning them into its political 'clients'. This ultimately led to an increase in the role of the centre in various spheres of centre-regional relations. Of particular importance were the system of fiscal and inter-budgetary affairs, regional cadre management, and the provision of social services in the regions. However, the overt reliance of the polity on the disposition of political forces at the centre and on the economic resilience of the federal budget brings into question the stability of the new regional system.

Indeed, the predominance of the centre in the redistribution of political powers and economic subsidies among the regions impeded the establishment of a system of incentives which would encourage central and regional political societies to maintain their dialogue regardless of the external pressures and the composition of ruling political circles. Therefore, regional loyalty to the centre is ensured by central economic concessions and administrative fear, thus under-mining the chance of establishing an independent and durable framework of federal relations. Furthermore, the success of these reforms strongly correlated with the popular acceptance of Putin's ideas and his high political rating. As Joel Ostrow (2002, p. 57) notes, 'personalised politics are inherently unstable. When support swirls around a specific person, as was the case with Yeltsin and now with Putin, it can easily fall apart depending on circumstances and the leader's status'. Finally, perpetuation of the most important legal and financial asymmetries, which originated from various forms of concessions to differing regions, put the effectiveness of the functional parameters of the new system and the feasibility of the conducted changes into question.

In the light of these ideas, this chapter will first evaluate the existing structural changes by examining the administrative, political, and economic resources of central and regional actors, as well as their strategies for achieving their respective policy ends. The discussion will then analyse the functional para-meters of the new regional system and the real political processes taking place within it. Above all, we will determine how far the Putin reforms have managed to alter the balance of power between the centre and the regions that became established during the Yeltsin era. The analysis will concentrate on the spheres of centre-regional relations that were subject to fundamental changes and the areas in which the centre and the regions managed to reach a reasonable compromise.

Central power resources and structural moves

The centre adopted a range of structural changes that enabled the Kremlin to expand its levers of influence over the regions on many important fronts. The new structure represented a comprehensive break with the Yeltsin era. Yeltsin's traditional methods of exerting authority over the regions were rather limited and included the federal monopoly on transportation systems, energy supplies, and provision of important financial services – Central Bank, the Ministry of Finance, *Sberbank* (Russia's main savings bank), the tax police, customs. This range of power instruments has been substantially expanded by the five following resource groups: (1) restrictive system of inter-budgetary relations; (2) the transfer of the right to control appointments in regional branches of federal institutions back to federal jurisdiction; (3) the right to remove regional governors and dissolve regional legislatures; (4) parliamentary restrictions; (5) pluralisation of regional political actors via the legal harmonisation process and enactment of new, democratising legislation.

First, Putin and his team of reformers regarded the central predominance in the sphere of budgetary and fiscal affairs as one of the key methods of exerting the influence of the federal centre over the regions. In particular, the introduction of a new Tax Law on 19 July 2000 (enacted on 1 January 2001) modified the redistribution principle for federal tax revenue in favour of the centre. During the Yeltsin era, the centre and the regions divided all the shared taxes on a 50/50 basis; the new government adjusted this balance towards a 60/40 principle (Khristenko 2002, p. 13). In addition to these changes, the centre introduced a business-friendly fiscal policy, which led to further decreases in the regions' tax base. From 1 January 2001 income tax was reduced to a flat 13 per cent rate;[1] small businesses with up to 20 staff and an annual turnover of up to 10 million roubles were given the choice of contributing either a 20 per cent tax on profits or an 8 per cent tax on revenues as of the start of 2003.[2] The 2004 tax policy assumed further decreases in VAT, which was lowered from 20 to 18 per cent, and the total abolition of the tax on sales, which was collected solely by the regional budgets. In addition, during the two years of tax reform between 2001–2, the government abolished a number of taxes designated solely for regional budgets, thus considerably reducing the fiscal base of the subjects of the federation: tobacco and road taxes, for example, were transferred to the jurisdiction of the federal centre (Belousov 2002). Similarly, in 2003, control over 50 per cent of the regional share of excise on petroleum products, which was used by the regions for road reconstruction and development, was centralised in the hands of the federal centre with the aim of a 'fair redistribution'.[3] By 2004, this policy deprived regional budgets of some 62 billion roubles of annual income while the compensation provided comprised just 21 billion.[4]

These developments have constricted the ability of the governors to control the political situation in their regions by traditional economic means. Because the governors were responsible for the provision of social services, transportation, education, and pensions and wages, the significant reduction of the regional fiscal base led the heads of the regional executives to rely on the additional transfers from the federal centre. Naturally, the effective delivery of such transfers was conditional on a demonstration of loyalty and political support for the policies of the centre. The Kremlin's growing reliance on such methods was graphically illustrated by the increase of federal subsidies to the regions during Putin's first presidency. While in 2002 the regions received just 200 billion roubles, the 2003 budget spared 700 billion roubles for similar purposes.[5] The 2004 budget continued this line by offering some 813.97 billion roubles for regional subsidies.[6] With the increase of gubernatorial responsibility in the social sphere, which often results in spending nearly 50–60 per cent of the entire regional budgets just on wages without accounting for any other developmental and social issues (Agaptsov 2003; Agaptsov 2001), such financing interlocked regional leaders with the centre politically and economically.

In the view of a number of governors, this incongruity between central budgetary and economic policies and regional abilities to comply with these requirements led presidential representatives in the regions to interfere in vital regional political and economic activities. If the governors failed to implement the economic demands of the centre in social policy, they would have to resort to the assistance of plenipotentiary representatives (*polpredy*) as the latter represented the federal centre. This in turn resulted in the *polpredy* 'recommending' to governors how and where to govern. In the Central Federal District, for example, this situation was an immense irritant to the governors of Kostroma and Moskva Oblasts, who pointed out the incompetence of their presidential representative on certain industrial and agricultural issues.[7]

These budgetary innovations were added to a number of the already existing means which enabled the centre to exercise tight financial control over the regions. A case in point is that the government and legislature draft the national budget from 'top to bottom' – a principle that enables the centre to evaluate the size of the slice that the regions receive from the federal budget. In this system servicing of the external debt, the needs of the army, and foreign and defence policy is given higher priority than financial support for the regions and municipalities (Khristenko 2002, p. 19). This system enabled the centre to pursue populist rhetoric partly through usage of regional economic resources and at the same time to control the rating of regional governors through distribution of federal transfers. This was particularly true during the early reform stages, when the government sought to restore Russia's international prestige by early repayment of external debt to international financial agencies and other nations – a policy that favourably contrasted Putin's performance with that of his predecessor and was positively accepted by the population.[8] This step led to a genuine resolution of the country's debt burden: by 2002, external debt had been reduced by $22 billion from $160 billion to $138 billion,[9] and at the end of 2003 Putin triumphantly noted that the country had successfully passed the peak of its international debt repayment by transferring some $17 billion to creditors without 'any visible strain or notice'.[10] The country's economic status within the world's leading credit and financial rating agencies was also steadily improving, which finally resulted in the achievement of an investment rating from Moody's in October 2003. Fitch followed this decision in November 2004, and Standard and Poor began to regard Russia as an investment-worthy country in January 2005. Undoubtedly, various external economic factors were responsible for this achievement. Nevertheless, this situation also became possible partly because the centre decreased the regional tax share and shifted a large amount of responsibility in the social sphere onto regional executives.[11] By doing so, the centre obtained significant resources to concentrate on these pressing issues of national importance. At the same time, however, Moscow placed overt pressure on the regional governments to fulfil their social obligations to their respective populations vis-à-vis the repayment of wages, pensions, and grants. In other words, the regions contributed considerably to the increase of central prestige and made national politicians ever more popular.

Second, the centre endeavoured to regain the right to influence appointments in regional branches of federal institutions, in particular in the legal enforcement agencies. The new policy aimed to turn the balance of power firmly towards the centre by attempting to ensure that the army, police, prosecutor's office, tax, and customs police is comprised of personnel loyal to the Kremlin and directly responsible to the presidential administration and its respective offices in the regions. Putin's decree No. 849 issued on 13 May 2000 stated that plenipotentiary representatives would thereafter take part in the selection of cadres for federal branches of legal enforcement institutions. Specifically, the document stated that the presidential envoys' duties would include 'implementation of the president's cadre policy… in federal institutions, where the president, government, or federal executives are entitled to make appointments'.[12]

The new edition of the law 'On the Militia', enacted on 4 August 2001, complemented this decree. The existing amendments were made in order to replace the old elite duo of 'governor–head of the regional MVD' with a new elite line: '*polpred*–head of the regional MVD–head of the territorial (Federal District) MVD' (Mikheev 2001). In accordance with the amendments, regional MVD ministries are nominated by the president following an introduction by the RF MVD minister; the regional governor has only a consultative function in this process.[13] The previous version of the law, on the other hand, demanded this nomination be made by the RF MVD minister on prior agreement with a regional governor.[14] The head of the District MVD branch is also appointed by the president, and this new institution comprises around 150 people in every District who work independently of their colleagues in the regions. The new amendments also abolished the procedure of negotiating the candidacy of local heads of regional MVD committees with regional leaders: the new law stipulates that the head of the regional MVD – who was after all appointed by the president – makes this cadre decision solely and independently.[15] The law was partially relaxed in 2005 – a development that we will discuss below.

Third, the centre attempted to move towards an explicit institutional hierarchy with Putin's May 2000 decision, which granted the president the right to unilaterally initiate procedures for dismissing the governors and dissolving the regional legislatures.[16] Despite that the subsequent Constitutional Court decision from 15 April 2002 made the procedure of gubernatorial removal from office contingent on judicial review, the opportunity for a federation president to eject a popularly elected governor from office remained in place.[17] The centre has particularly accentuated its right to initiate criminal proceedings against the governors and heads of regional legislatures if they violated federal law. Putin's original proposal seemed harsh and evoked concerns because of the possibility of bypassing the courts to reach a decision on the suspension of a governor and deployment of the criminal investigation procedures in this process. Figure 3.1

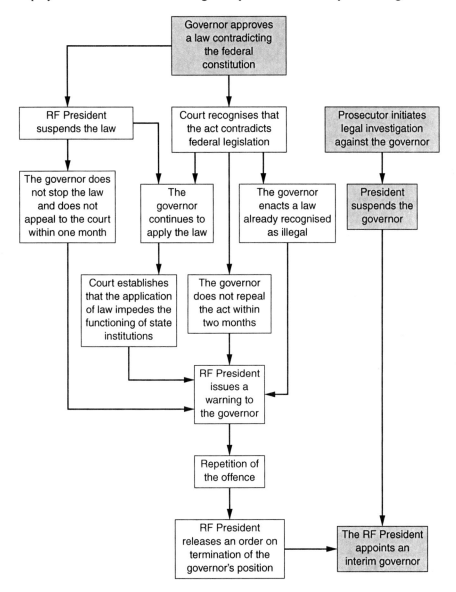

Figure 3.1 Sequence of events in the case of presidential dismissal of governors.[18]

illustrates the actions required to dismiss a governor in the initial presidential plan.

The Kremlin severed this arrangement in September 2004 by introducing the system of electing the governors by the regional assemblies on the introduction of the president. The new model provisioned that, instead of a direct popular vote for gubernatorial offices, a selection of candidates must be submitted to the president by plenipotentiary representatives in the Federal Districts. The president would then have to choose one candidate and submit this nominee to a vote by regional assemblies. The head of state has also received the right to disband the assembly and nominate an interim governor should a legislature refuse the presidential nominee twice. The governors have been given the opportunity to avoid this procedure by resigning voluntarily and appealing directly to the president for a vote of confidence. The legislative package accommodating the new appointment system included Federal Law No. 159-FZ, which was passed by the State Duma on 3 December 2004 and signed by the president on 12 December, and presidential decree No. 1603, issued on 27 December 2004. The first document amended the laws on 'organisational principles of executive and legislative bodies of the Russian Federation's subjects' and 'on main guarantees of the electoral rights and participation of citizens of the Russian Federation in a referendum' and stipulated new procedures for appointing and removing governors from office. The second document outlined the procedure for the selection of candidates by the plenipotentiary representatives for the presidential nomination.[19]

Fourth, the centre obtained significant power resources within the parliamentary sphere. This has had a largely negative impact on the evolution of centre-regional relations. The May 2000 reform of the Federation Council (FC) drastically weakened the political weight of the upper chamber. The policies substituted the previous *ex-officio* system with appointing delegates chosen by the regional administrations and legislatures.[20] In these settings, the upper chamber has effectively ceased its independent functioning as a chamber of regions. Seventy-five to eighty per cent of the new nominations were either recommended by or cleared with the presidential administration (Remington 2003, p. 674), while as many as 34 per cent of the new Council members had direct links with Russia's large and medium-sized corporations.[21]

It is also significant that the regions have become precluded from effectively representing their organised interests during the lower house legislative process. The December 2003 victory of the pro-presidential *Edinaya Rossiya* party granted this organisation control over the key Duma positions and restricted the factions' formation rules, thus targeting the pre-existing regional formations.[22] Similarly, responsibilities for drafting important budgetary proposals have been transferred under the jurisdiction of federal economic ministries. This prevented the regions from voicing their opinions during the initial legislative stages (Chirikova and Lapina 2004). The December 2007 enactment of the proportional representation system further destroyed the regional parliamentary strongholds traditionally represented by plurality mandates.[23] Turovsky (2007b, pp. 80–2) demonstrates that the use of single mandates in the interests of capital elites was minimal, and the share of Muscovites running for the SMD seats almost negligible during both the 1999 and 2003 elections. This contrasted the highly centralised nature of the party lists seats redistribution, in which the majority of the mandates were transferred to representatives of Moscow elites. *Edinaya Rossiya*, which along with the Communists represents the most genuine, and thus regionally entrenched, national party in Russia, transferred 54 of its 120 December 2003 mandates to its capital bosses.

Fifth, the centre made an attempt at pluralisation of political actors in the regions and removing total political control from the hands of the governors. As opposed to the other moves discussed, this range of initiatives did not have an overtly hierarchical, constraining characher. For many such policies were achieved through the legal harmonisation process and through the issuance of legislation aimed at democratising regional political climate. The new institutional system demanded that the regions comply with the single standard for federal and regional legislation, and made regional leaders and parliamentarians directly responsible for drafting laws found to be incompatible with the federal constitution. Putin's presidential decree No. 849 issued on 13 May 2000 divided the country into seven federal districts headed by plenipotentiary representatives who were simultaneously members of the presidential administration. Apart from general co-ordinating and supervisory duties, one main function of the *polpredy* was to ensure that the regions do not issue legislation that violates constitutional norms and principles of the Russian Federation. These measures immediately produced an effect: as early as April 2001, most violations of the federal constitution and federal law had been addressed and four-fifths of regional legislation had been brought into compliance.[24]

Furthermore, the changes increased the influence and political importance of regional legislatures as the latter could advocate the national parties' interests in the regions, assist regional co-operation with the centre and to a certain extent contain previously unlimited powers of the regional leaders. Regional legislatures gained the right to approve the regions' budgets and control their implementation – a practice that never existed during the Yeltsin period. In addition, in summer 2002, the centre demanded an increase in the number of deputies employed in the regional legislatures on a full-time basis and insisted that at least 50 per cent of all deputies in regional legislatures must be elected on the party list electoral principle.[25] Inconsistencies related to the principle of separation of powers on regional and municipal levels were removed with the aim of liberalising the functioning of regional parliaments. On 25 July 2002, a law was enacted which prohibited combining the position of people's deputy in the regional legislatures with executive posts in municipal structures, i.e. mayors.[26] These developments, though partially restricting the authority of the governors and mayors, represented a positive development towards refining democratic principles of governing within the regions and between the regions and the centre.[27]

More importantly, a number of laws were introduced to facilitate the process of concluding bilateral agreements, to make it more open and legitimate, and to protect such bills from being perceived as a political irritant. In his 2002 Address to the Federal Assembly, Putin emphasised that radical changes had taken place in the process of concluding the treaties. In particular, he insisted that 'it is unacceptable to conclude bilateral agreements behind the back of other subjects of the federation without extensive discussions and the development of an outright public consensus. All possible bilateral treaties should go through the procedure of ratification by the parliament so as to make transparent which privileges one or another subject receives and why'.[28] Such a declaration was preceded by serious legislative work: on 24 June 1999, the legislature adopted a law 'on the principles and order of division of responsibilities between the administrative bodies of the federal centre and subjects of the federation'.[29] The law demanded discussion of texts of the newly established treaties in regional legislatures (a practice which was not usual prior to the enactment of this law) and passing these texts further to the Federation Council for final approval – a practice which had never previously existed.[30] Only then could the president and the head of a region bilaterally sign a newly formed treaty. The law has also obliged the subjects of the federation to bring the existing treaties in line with the acting constitution; a failure to comply with this requirement automatically annulled the power of these treaties.[31] These measures largely resemble the already discussed bargaining practice established in Spain, where reiterative centre-regional power negotiations represent an objective political reality and could not be avoided. More importantly, this set of actions, whilst preserving the treaties as a possible form of legal relations between the centre and the regions, considerably diminished the window of opportunity for regional 'sovereignisation' by means of the Federation Treaty and bilateral treaties, which in itself represented a reasonable compromise between the two major actors, the regions and the centre.

Regional responses and federal processes

As we have already mentioned, the real federal processes resisted the centre's structural drive towards centralisation and managed to create a system of adaptive federal relations, in which the regions retained a substantial scope for manoeuvre. As a result, a number of key areas, which most needed the reform, remained intact. Moreover, due to the apparent 'elite settlement', the regions have begun to display a substantial degree of conformism, the willingness to pursue their political goals via non-transparent shadow negotiations, and the tendency towards accumulating feelings of hidden resentment against the federal centre. Thus, the texture of regional politics, as well as the complexity of the pre-existing centre-regional relations, has not been substantially altered, despite the Kremlin's structural efforts to change the situation. Let us trace these dynamics with the example of some key spheres of the regional dialogue.

Inter-budgetary relations

The new system of centre-regional budgetary relations has largely led to the acceleration of informal centre-regional negotiations and bargaining. Despite central initiatives to launch the drive against so-called 'budgetary separatism' by introducing the system of inter-budgetary relations based on single norms and principles, a number of financial asymmetries, in particular those concerning republican privileges, remained largely intact. The new arrangement yet again represented a system which relies on offering economic incentives to selective regions with the aim of persuading them to remain loyal to the federal centre. The new system differed from Yeltsin's arrangement in the sense that it placed greater emphasis on tighter central control over the regions' financial activities. In all other aspects, many analysts and politicians have often insisted that, in the sphere of inter-budgetary relations, Putin's team 'was steadily repeating Yeltsin's path'.[32] Comparative analysis between Putin's and Yeltsin's frameworks of centre-regional financial interaction demonstrates that the two systems did not differ markedly in the sense that they both attempted to glue the subjects of the federation together by means of various economic incentives and political concessions. In Yeltsin's case, the system provided the regions, and in particular the republics, with selective tax exemptions. During the 1990s, often referred to in political literature as a period of 'budget separatism', Tatarstan – along with other pioneers of politico-economic sovereignty such as Bashkortostan, Yakutiya and Chuvashiya – did not contribute taxes to the federal budget. This situation did not change until 1996. These areas retained various fiscal privileges thereafter allowing just Tatarstan to migrate from 43rd to 23rd place in the level of socio-economic development among the federation subjects (Borodai 2002). If Tatarstan transferred taxes to the federal budget in accordance with common rules applied to all the other regions, the region would loose some 16 billion roubles – a third of the entire regional budget!

Putin's policy towards reforming this fiscal situation failed to alter the essence of the system. On the surface, most 'chosen' regions were stripped of all official privileges with the theoretical aim of equalising the federation subjects in their tax duties. At the same time, the government replaced virtually the whole share of these regions' 'financial losses' with specialised, extensive federal subsidies. In the case of Tatarstan, on 12 June 2001 the central government approved the federal programme for 'Socio-economic development of the republic of Tatarstan until 2006', which assumed large investments by business into the republican social sphere. The figure for these investments virtually compensated republican losses incurred by the newly introduced regime of equal fiscal duties (Borodai 2002).[33] Moreover, even after the enactment of the reforms Tatarstan continued to make minimal contributions to the federal budget. In 2000, the republic transferred no more than 15 per cent of collected taxes; in 2001, despite the new 40–60 fiscal share between the centre and the subjects, the region transferred just 15.9 billion roubles to the centre whilst retaining 20.5 billion in the republican budget.[34] Finally, even in 2003, when regional opposition became less prominent, the centre provisioned the most impressive federal transfers from the 2004 budget not for the poorest regions but for relatively rich Tatarstan, Bashkortostan, and Saratov Oblast.[35] A similar situation took place over resource-rich and financially stable regions of the Russian Federation. Various donor-regions retained markedly different parts of their income within their territories, depending on their political importance and personal connections within the Kremlin. In 2006, Khanty-Mansi Autonomous Okrug managed to keep just 17 per cent of its revenues in the regional budget, while Moscow was able to retain as much as 59 per cent. St Petersburg, as a particularly well-connected federation subject, retained 67 per cent of its income in the regional budget.[36] It is also important that by 2007, 61 per cent of all the existing federal transfers to recipient regions had been channelled on the basis of shadow negotiations, in which political factors, such as the amount of votes given for the pro-Kremlin *Edinaya Rossiya* party, became defining. Thus, many regions received federal assistance not on the basis of socio-economic needs but as a result of their political status and personal connections at the federal centre (Zubarevich 2007).

In the investment area, the Kremlin recognised the need for immediate action within the regional dimension, if it were to fulfil the task of the country's economic revival. Yet again, the level of direct investments was largely dependent on the regional administrations' personal connections at the federal centre. Turovsky (2007a) insists that the regional economic initiative did not play any significant role in attracting investors, and given that all key decisions were taken at the federal level, the Kremlin was responsible for directing investors into politically loyal territories. Within the public dimension, however, the centre pledged to select territories of high developmental potential and target its financial currents into these areas on an equal basis. This primarily concerned the launch of the Investment Fund and the establishment of the 'special economic zones'.

The 'special economic zones' project was launched in 2005, with three types of such zones being distinguished: industrial, innovative, and recreational. These zones, which provide substantial administrative and fiscal relief, can be established in any federation subject with the aim of attracting investment, giving a push start to the development of poor territories, and thereby levelling the existing regional disparities. However, despite these declarations, Russia's politically important territories topped the short lists of regions selected to host such zones. By 2008, 13 special economic zones had been created in Tatarstan, Moscow region, St Petersburg, as well as Tomsk, and politically loyal Lipetsk oblasts.[37]

The state also established a special Federal Investment Fund intended to support substantial regional development projects in 2005. By 2007, the Fund had accumulated some 265 billion roubles of currency reserves. At the same time, business representatives admit that, while it is almost impossible to push an economically unviable project through the Fund's rigorous assessment process, obtaining grants from this institution is only possible via intensive lobbying and corruption (Shmarov and Stolyarov 2007). The government was also hesitant in utilising the Investment Fund resources for regional developmental needs, thus creating the problem of asset depreciation under inflationary pressures. This was related to the fact that the funds redistribution process was overwhelmingly informal and the state had no transparent criteria for allowing the regions to benefit on an equal basis. In 2008, the new head of the Ministry of Regional Development Dmitry Kozak emphasised the need for developing a coherent legislative basis for funds' redistribution and called for relaxing the existent funding eligibility criteria.[38] The cabinet supported this initiative by proposing to co-invest various regional infrastructural projects on a non-refundable basis.[39] At the same time, the prevalence of informal means of centre-regional political inter-exchange made the state lose nearly three years of potential utilisation of the Fund's resources.

Such tactics and policies distanced Russia from the adoption of a solid system of incentives for regional politicians to become part of the federal political elite, and resolve the existing socio-economic problems in coalition with the centre. The existing arrangement, while making the centre-regional financial relations appear more equitable on the surface, is invariably unstable and holds its future a hostage to the economic policy of the centre and national macroeconomic indicators. This sharply contrasted with the US system, which largely draws its success from the fact that the founders of the country's institutional design and constitution did not rely on the promise of prosperity for the viability of their confederation, but in turn considered circumstances that were threats to prosperity (Ordeshook and Shvetsova 1997, pp. 31–3).

The shaky nature of this system had become practically visible by 2005. By then, a number of governors had embarked on collective protests that targeted certain economic policies of the federal centre. These protests were very short-lived but had the ability to demonstrate that the regions have begun resenting many central initiatives. For example, in the wake of the 2005 monetisation of state benefits, 13 governors, members of the Far East inter-regional association, forwarded an open letter to the president claiming that the monetisation programme violated the Russian constitution and denied the country a genuinely federal institutional structure. Similar political struggles occurred over the 'two keys' licensing principle. The 'two keys' rule, embedded in the 3 March 1995 version of Law 27-FZ 'On Natural Resources', implied that the regional leaders' signatures were required on all mining licenses. This granted the governors the 'second key' to Russia's natural resources. Article 13 of Federal Law 122-FZ, adopted on 22 August 2004 and enacted on 1 January 2005, abolished this rule and effectively transferred control over the country's natural resources solely to the jurisdiction of the federal centre.[40] In response to the 'two keys' abolition, the Mayors of Moscow and St Petersburg Yurii Luzhkov and Valentina Matvienko, as well as the Presidents of the republics Tatarstan, Bashkortostan and Yakutiya, Mentimer Shaimiev, Murtaza Rakhimov and Vyacheslav Shtyrov respectively forwarded a note of protest to the president.[41]

Silent non-compliance and pursuit of regional policy against the wishes of the federal centre has become another most important manifestation of the regional drive towards autonomy. The centre, for example, has turned a blind eye to the fact that in 2005 one fifth of Russia's regions decided against implementing the monetisation initiative. Among the most defiant territories were Russia's richest subjects: Moscow Federal City, Tatarstan, Khanty-Mansi Autonomous Okrug, Krasnoyarsk Krai, as well as Kemerovo and Sverdlovsk Oblasts.[42] More importantly, these regions found quiet ways of doing so by redistributing federal grants only and simultaneously maintaining the ongoing benefits system.

Finally, the regions have pressed for a looser union within the taxation sphere. Given that the new fiscal redistribution system had deprived the subjects of 62 billion of annual income by 2004,[43] the latter have embarked on adaptive institutional strategies for defending their interests. Pressure duties have often been delegated to the regional assemblies, given their de-facto independence from the Kremlin. The most recent initiative emanated from a group of Saratov legislatures who, supported by their colleagues from Nizhnii Novgorod, Samara, and Ulyanovsk, forwarded an open letter to the then Prime Minister Fradkov demanding the revision of the current principles of fiscal redistribution (Gazizova 2007a). Partly in response to these demands, the centre has returned income tax, as well as corporate property and land taxes, to the regional jurisdiction.

Appointment policy

Despite the introduction of formal rules for appointment mechanisms, it is doubtful that the centre was entirely successful in implementing these innovations in practice. The Kremlin was unable to exclude the governors from this sphere entirely, leaving the latter substantial room for manoeuvre. The first term of Putin's presidency was marked by covert gubernatorial pressures for the nomination of their favoured candidates, while the second term witnessed the legal retreat by the federal centre coupled with the accelerated activity of regional leaders in the appointment sphere. In the aftermath of rigorous fieldwork, Russia's leading regional expert Chirikova (2003, p. 102) concludes that, between 2000 and 2004, the governors continued to control important appointments within regional branches of federal institutions. This primarily concerned police, sanitary inspection institutions, customs and tariffs, fiscal, ecological, and energy structures. Some of these bodies were still financed by the local authorities and therefore their leaders remained loyal to the regional governors.

Indeed, until 2005, appointments in territorial branches of federal institutions of many regions largely resembled the old practice, taking place within a framework of informal agreements between presidential representatives, heads of regional administrations, and respective central ministries. A number of regional leaders managed to find ways around the new system and to retain some influence in the appointment sphere. A conflict in Moscow in winter 2001, when Mayor Luzhkov called into doubt the nomination of the presidential representative's protégé General Shvydkin to head the regional MVD branch, illustrates this point. Luzhkov went far to demonstrate his power potential: he made numerous appearances on the city's TV and radio channels and demanded that *polpred* Poltavchenko speak to the city Soviet with regards to Shvydkin's nomination.[44]

Luzhkov's struggle in Moscow had repercussions in other regions. Governor Rossel of Sverdlovsk Oblast created a special appointments mechanism, which ensured that unsuitable nominees would not appear in important regional institutions. He negotiated with some federal ministries such as the State Pension Fund and the State Property Committee that the presidential representative would not make any appointment without the prior agreement of the regional administration.[45] Officials in Moscow region did not seem concerned about the situation either: deputy governor Nikolai Rebchenko insisted in 2002 that, despite the new version of the law 'On the Militia', appointments in legal enforcement agencies would be made following mutual – mostly private and non-official – agreements between the presidential envoy and the regional administration.[46]

This situation underscored Russia's reliance on her informal and arbitrary political style. On the structural front, however, it came about for the following two reasons. First, as Badovskii (2001) insists, the number of districts created is too small to produce an immediate effect. All previous regional branches of federal institutions remained in place, i.e. city, *raion*, and *oblast* offices, which made it difficult to associate such a large number of local institutions with a federal district's centre and therefore perpetuated the established appointment culture. Second, as Smirnyagin (2001; see also Zubarevich *et al.* 2001, pp. 185–90.) often mentioned, the borders of some of the institutional districts such as the MVD and judiciary should not converge with the borders of the federal districts. By virtue of a simple increase in a number of participating actors, this system would help to reduce corruption and familiarity in the appointment process, and make the system impartial, competitive, and effective. Unfortunately, Putin's reform has ignored this important requirement. The reform of the MVD has based its institutional districts on the federal ones, and the country's military districts, which inspired the emergence of the federal districts, had already converged with them.[47] The subsequent creation of various administrative districts for various federal agencies – the State Building and Construction Agency (2003), the Federal Service for Control of the Circulation of Narcotics (2003), the Federal Financial Monitoring Service (2003), and even the Russian Academy of Science (2002) – has also followed the established federal districts pattern (Petrov 2006, pp. 78–9).

The apparent central inability to effect serious changes within the appointment sphere have led to the revision of the previously adopted legal code. Given that many governors lamented their inability to execute over two thirds of their current responsibilities, the 2 July 2005 State Council meeting decided to return to the regions the right to influence nominations within the regional branches of federal institutions. This decision was formalised with Putin's decree No. 773 of 2 July 2005.[48] The decree enabled the regional leaders to organise and co-ordinate the functioning of the territorial branches of federal ministries: the Ministry of the Interior, the Ministry of Justice, the Ministry of the Emergencies, the Federal Penal Service, the Civil Defence Service, the Federal Registration Service. The federal ministries were obliged to obtain a regional governor's agreement upon their selected nominations to executive positions within these institutions, and to offer an alternative candidate in case of refusal. The nomination, however, could still take place should the governor refuse a proposed candidate twice.

It is also important that, despite these changes, regional governors were frequently involved in political conflicts with the heads of territorial branches of federal institutions. Very often, such conflicts were accompanied by gubernatorial demands to remove these executives from office. In summer 2007, for example, the Khakassiya regional Ministry of Interior was engaged in a far-reaching conflict with the regional authorities. Governor Alexei Lebed and the head of the regional legislature Vladimir Shtygashev forwarded a note of complaint to the federal centre with demands to effect relevant personnel changes. Similarly, in the Kemerovo region the authorities struggled against the Technical Supervisory Service over the problems of industrial safety regulations in the regional mining shafts (Kynev 2008).

Bilateral treaties

The victory of the federal centre on the bilateral treaty issue was also rather limited. On the one hand, the treaties' practical usage diminished markedly, and some of the agreements – in particular those between non-ethnic regions and the centre – were renounced. According to Russian government data, between 2000 and 2002, 28 out of 42 subjects of the federation voluntarily withdrew from these agreements in the framework of aligning their legislation with federal standards.[49] On the other hand, the drive to renounce the treaties originated from those regions that were the last to sign the documents and which received the least possible benefits from the process.[50] The first moves towards cancellation of these treaties were made by the governors of four such regions in July 2001. Yurii Trutnev of Perm Oblast, Vladimir Shamanov of Ulyanovsk Oblast, Ivan Sklyarov of Nizhnii Novgorod Oblast, and the President of the republic Marii El Leonid Markelov launched simultaneous claims to renounce their treaties with the federal centre.

Therefore, despite the fact that some 'decorative' agreements were renounced, those documents concluded via long processes of negotiations – such as that with Tatarstan and Bashkortostan – survived. These regions still rely on bilateral agreements as a means of communication with the federal centre. For example, the Republic of Tatarstan legitimised the treaty in the revised version of the republican constitution, making it – along with the Constitution of the Russian Federation and the Constitution of Tatarstan – a legal basis of the republican association with the Russian Federation.[51] Bashkortostan has also retained the reference to the bilateral treaty in its basic law.[52] Similarly, treaties concluded with a number of politically important regions such as Sverdlovsk, Kaliningrad, and Irkutsk Oblasts, as well as with Moscow and St. Petersburg federal cities, remained in place until their respective expiration dates.

A certain revival of the treaty conclusion process took place in 2006–7. Tatarstan and Chechnya were at the forefront of these dynamics. As the republics did not sign the 31 March 1992 Federation Treaty and did not vote in the 1993 Russian Federation constitution (Kahn 2002, pp. 151–7), signing bilateral agreements with both these regions represented a crucial step towards instituting their membership in the federation. Tatarstan deployed this situation, along with the fact that the original 1994 Russo-Tatar Treaty lost its force in 2003. Thus, the bargaining process in the bilateral agreement sphere never ceased. Notwithstanding a wide range of Kremlin-inspired institutional complications,[53] the new Treaty was signed into law in July 2007.[54]

The document granted the region some serious concessions. First, the republic legitimised its treaty-based, negotiated relations with the federal centre. This reinforced the republican constitution, which as we have already observed, bases Tatarstan's association with Russia on three pillars: the Federal and Republican Constitutions and the bilateral treaty. Moreover, the Treaty was ratified by regional and federal parliaments. It did not contradict the national constitution and received the status of federal law. This made the legal and political relations between the two parties more transparent.[55] Second, Tatarstan received certain republican citizenship rights. The Treaty allowed its residents to carry specific nationality slips in their passports written in the Tatar language and containing the republic's symbols.[56] The republican authorities were also encouraged to promote the Tatar culture outside the region's borders. Third, Article 91 of the 1992 Tatar constitution, which demanded all presidential candidates be fluent in both Russian and Tatar, was reinstated, despite the initial abolition by the federal centre. Finally, the Republic received substantial concessions in the spheres of mining taxation and international political and economic activity (Gazizova 2007b).

The centre has also pledged to provide Chechnya with the 'widest possible' autonomy. The republic has modelled its draft Treaty on the 1994 Russian-Tatar agreement and demanded a unilateral access to natural resources, substantial tax relief, and the imposition of restrictions on the dislocation of federal military.[57] In June 2007 the republic officially requested the Kremlin to create an offshore economic zone that would involve a 14-year federal tax relief and substantial duties relief for goods produced within the region.[58] At the time of writing, the treaty negotiations were put on hold by the mutual Moscow-Chechen arrangement. However, there are reasons to believe that the process will be reinstated in the near future.

The treaty negotiations were not confined to Tatarstan and Chechnya. In the immediate aftermath of the Tatar Treaty annunciation, the head of the Bashkir parliament Konstantin Tolkachev declared Bashkortostan's determination to sign a similar agreement. In this case, further pressure duties have also been delegated to the republican parties. The latter established a movement called For Federal Russia. The organisation pledged to mobilise the regions willing to conclude similar treaties and to promote a genuinely federal structure of Russia's centre-regional relations.[59]

Legislative compromises

The legislative area of centre-regional relations has also witnessed a number of important compromises. President Putin employed his right – provisioned in Article 85.2 of the national constitution – to suspend regional legislation that contradicts federal standards and involves violation of human rights across the country. In 2000 he issued 18 decrees putting a ban on legislation in differing regions which concerned agricultural and transportation matters, migration, trade, licensing, and tariffs (Vasileva 2001, p. 18). By April 2001 Putin had proudly announced that most violations of the federal constitution and federal law had been addressed.[60] However, the grounds for such optimism were rather tenuous. As Hahn (2003, pp. 147–9) points out, legal harmonisation represents an unremitting process even in the most successful federations and, despite some positive developments, generates a range of important problems in Russia. Indeed, the regions continued to violate federal law and, by 2004, complaints by the plenipotentiary representatives on difficulties associated with legal harmonisation had become a distinctive trend.

In 2004 the former presidential envoy to the Southern Federal District Dmitrii Kozak insisted that statutes and laws in the District's regions contained a vast number of contradictions of federal legislation, and he threatened to dissolve regional assemblies if the latter did not make greater efforts to comply with the national standard.[61] Similarly, despite the fact that as many as 4,138 legal acts had been amended over the course of four years in the North-West Federal District, a large proportion of the legislation still contravened federal norms. In May 2004 *polpred* Ilya Klebanov pointed out that the Republic of Kareliya had 25 violations of the federal law in its statutes, the Republic of Komi – 18, Arkhangelsk region – 20, Vologda Oblast – 29, Kaliningrad region – 24, Murmansk region – 13, and Pskov Oblast – 17.[62] Similarly, only 625 of the 1,500 legal acts violating the federal standard in the Far East Federal District had been brought into compliance by February 2003. The remaining 875 provisions were legally contested by the District's regions.[63] Violations of federal legislation in the Ural Federal District grew by 28 per cent during 2002. Deputy General Prosecutor in the District Oleg Gorbunov insisted that such an increase had taken place mainly because of the policies of the regional authorities, who often pursued private goals at the expense of national interests.[64] In the Central Federal District, despite efforts at harmonisation, the *polpred*'s team discovered over 12,000 local legal acts that contravened federal legislation in 2002 (Poltavchenko 2003).

The national republics represent a special case. The perpetuation of political and legal asymmetries, particularly in areas such as provision for republican sovereignties, citizenships, differing names and statuses of the federation subjects (the existing possibility of upgrading the statuses invariably leads to the perception of regions' differing constitutional positions) serves as a clear demonstration of the limitations of legal reforms.

Tatarstan, having accepted some 128 amendments to the 130 articles of its constitution,[65] still figured as a leader in the number of laws which did not comply with the federal standard. This was especially evident when the Republic decided to retain the clause on sovereignty in the new version of its constitution.[66] According to a number of commentators, the new definition of the republic represented a clear step forward in relation to the old one, which referred to Tatarstan as an 'independent state *within* the jurisdiction of the Russian Federation'.[67] However, by constituting state sovereignty in the new document, republican legislatures openly ignored binding presidential order No. 1486 from 10 August 2000 'On additional measures for the provision of the integrity of legal space in the Russian Federation'. This order refers directly to decision No. 92.0 of the Constitutional Court from 27 July 2000, which emphasises that the Constitution of the Russian Federation does not allow or recognise the state sovereignty of any of the federation subjects and clarifies that the multinational people of Russia represent the sole carrier of state sovereignty.[68] In addition, constitutional provisions, which state the existence of a separate republican citizenship constituted in republican passports, directly impede federal principles outlined in national legislation.[69] Moreover, despite the fact that Article 108 of Tatarstan's previous constitution – which required electoral candidates for the republican presidency to be Tatar-speaking citizens of the republic – was amended to establish greater equality among the citizens of the Russian Federation, Article 91.1 of the new constitution still demanded that the potential candidate should speak either of the republican state languages. These, according to the new constitution, are Russian and Tatar.[70] Similarly, in 2003 the Tatarstan prosecutor Kafil Amirov insisted on amending articles of the law 'on the languages of the people of the Republic of Tatarstan' regulating communication with foreign countries and international organisations and official inter-exchange in the spheres of transportation and telecommunications, and in the energy sector.[71]

Bashkortostan also resisted the centre's drive for legal harmonisation. The process of bringing the republican constitution into line with the federal standard took almost two years and involved numerous political conflicts and court trials. In December 2000 the Deputy General Prosecutor in the Volga Federal District Alexandr Zvyagintsev contested 55 provisions of the old republican constitution. Following a series of trials that were taken even to the national level, the higher court of Bashkortostan ruled against 20 articles of the basic law, essentially abolishing the clauses on sovereignty and renouncing the bilateral agreement between the centre and the republic as the basis for relations between the two parties.[72]

Despite these efforts, however, the new republican constitution was not a qualitative improvement of its predecessor. As Sakwa (2004, p. 156) notes, the new document was adopted in order to divert central pressure for legal harmonisation and 'violated federal norms as much as the old one'. Indeed, a range of provisions on local self-governance allowed the president to nominate the heads of municipal structures and included local institutions in the system of republican state authority.[73] The language requirement for the post of president was also retained in the new version of the document.[74] More importantly, the new constitution enabled the republican chief executive to nominate the heads of regional electoral commissions – an area that according to the law 'on main guarantees of electoral rights and participation of the Russian Federation citizens in a referendum' belongs to the federal centre.[75] It is also important to mention that the deputies of the republican State Council initiated criminal proceedings against the head of the republican Higher Court Marat Vakilov, who contributed most in the ruling against the previous version of the constitution.[76]

Such violations of the federal law were not unique to the ethnic republics: Moscow federal city is a case in point. As Petrov (2000b) insists, 'being the last to sign a bilateral agreement on the delineation of authority with the federal centre, Moscow had a status that was only slightly different from that of Tatarstan in terms of citizenship (*propiska*), control over law enforcement agencies and the courts, the lack of elected heads of local self-government, and special privatisation agreements'. Stavropol region was another leader in the number of acts contradicting federal legislation. The previous version of the regional Statutes stipulated various discriminative provisions towards non-residents, forbidding them from participating in local and regional elections and referenda and restricting their access to social services (Bartsis 2001). In June 2002 – despite adopting a range of amendments to the Statutes – the regional Duma passed a law on 'the measures for prevention of illegal migration to Stavropol region'. This document provisioned a range of restrictions on the number of people that were allowed to reside in various parts of the region, which contravened constitutional clauses on the free movement of citizens.[77] Despite repeated objections from Stavropol Krai's prosecutor, the regional legislature continued to apply the law until summer 2003, when the trials on its compliance took place in the Russian High Court. It is remarkable that the Court adopted a political decision to rule in favour of the region, notwithstanding the apparent violations of federal legislation.[78] This situation is not unusual for a number of other Russian territories, which are still 'closed' to migrants. In Krasnodar Krai, the situation over territorial migration was also particularly difficult.

Furthermore, a range of legislation issued by the federal centre with the aim of trimming back the influence of the regions on the national political landscape has been amended to incorporate a number of important compromises. The first term of Putin's presidency was particularly rich in such retreats. Even though with the enactment of the gubernatorial appointment system many of these legal compromises lost their former significance, it is important to examine them with the view of tracing the developmental dynamics of the centre-regional balance of power.

Most importantly, Putin's controversial proposal – to grant the federation's president the right to initiate procedures for dismissing the governors and dissolving regional legislatures in cases involving a direct violation of federal legislation – failed to come through in its original arbitrary form. Putin's initial legislative package – passed by the State Duma on 1 June 2000 with an overwhelming majority at the first reading – evoked concerns because of the possibility of bypassing the courts to reach a decision on the suspension of a governor (see Figure 3.1). Critics of this proposal were mostly apprehensive about the impact of the new order on some aspects of representative democracy: the very existence of a legal ability for the supreme executive to eject a democratically elected official from office would have undermined the principles of representation and federalism.[79]

This legal dispute was resolved by reaching a compromise in the Constitutional Court one and a half years later. Despite the fact that the Court hearings took as long as 18 months – many analysts speculated that the centre purposely put pressure on the Court to delay its decision – the final verdict appeared to restrict the president's ability to dismiss regional executives and dissolve legislatures. Whilst admitting the constitutionality of presidential proposals on the 'functioning principles of regional executive and legislative organs', the Court made a number of provisions practically forbidding the president from arbitrarily dismissing regional governors and dissolving regional legislatures on the basis of their issuing decrees contradicting federal legislation. The Court ruling underscored that the 'necessity of federal involvement cannot be justified solely by the fact of a formal contradiction between regional and federal acts'. In turn, the country's judiciary would have to establish that the application of such a law had led to 'mass and brutal violations of human rights and freedoms, as well as to a threat to territorial integrity, national security, defence, and the integrity of the legal and economic space of the Russian Federation'. More importantly, the Court ruled that if the president decided to initiate the suspension drive against the governor in practice, he should first obtain permission from the Constitutional Court. Even then, the Court must rule against the regional executive twice: first confirming the fact that the region adopted legislation contradicting the federal constitution, and second by stating that the regions' officials purposely avoided changes in the given law and by doing so knowingly violated the federal constitution.[80] At the same time, the Court confirmed the presidential right to suspend the governors following a request by the General Prosecutor on the basis of a criminal investigation related to severe offences. Even in this case, however, the governors retained the right to contest the president's decision in the High Court.[81] Such an extended procedure technically precluded the president from dismissing an inconvenient governor. Referring to the complexity of gubernatorial removal from office, Ross (2003, p. 40) quotes Corwin (2000) as saying: 'the process is so long and involved that regional leaders would have to demonstrate unprecedented obstinacy, audacity, and even stupidity before they could be fired'.

Putin's subsequent attempt in November 2002, to introduce a range of new legal provisions that would enable the centre to eject governors from office by using simpler procedures than those established by the Court, met severe resistance in the State Council. Therefore, the final edition of the law enacted on 4 July 2003 maintained all the Court's original provisions with regards to gubernatorial dismissal.[82] This law deserves special attention. Article 26.9 provides the federal centre with the legal opportunity to assume temporary control over the regions in three distinct cases: (1) natural disaster; (2) if regional debt to the federal budget exceeds 30 per cent of its total revenue; and (3) if regional executives violate federal law or the constitution while deploying specially designated federal subsidies.[83] Initially, these provisions seem restrictive. However, this law also represented an important compromise. First, the above provisions were supposed to be enacted only on 1 January 2007, thus providing a substantial political time delay for all the incumbent governors. Second, these provisions do not assume the suspension of governors and subsequent re-election. In particular, the introduction of external financial administration represents a temporary measure that could only last for up to one year or until a complete restructuring of regional budgetary and financial parameters. Moreover, such an administration cannot be enacted within one year of the election of the regional executive. Finally, all three cases assume judicial participation. In the first case, the president must obtain permission from the Federation Council to introduce external crisis management. In the second case, the federal centre has to secure a ruling by the federal Arbitration Court confirming the regional financial problems. In the third case, the procedures determining the violation of federal law and the constitution must abide by the 4 July 2002 ruling of the Constitutional Court, as discussed above.[84]

The cancellation of direct popular voting for the governors enacted in January 2005 has changed this situation drastically and largely devalued many of these compromises. At the same time, despite the enactment of such constrictive changes, the real fabric of centre-regional relations retained many elements of serious complexity. While appointing the governors, the centre placed a particular emphasis on the good management issue, which underscored the Kremlin's inability to dismiss the governors at will and pointed at the existence of some disguised forms of recognition of entities. The Kremlin, assisted by the Ministry of Economic Development, instituted a new gubernatorial effectiveness scale that could to some extent formalise centre-regional political negotiations within the appointment sphere. The assessment scale comprised 43 independent indicators within various socio-economic spheres.[85] These included regional GDP, industrial and agricultural indexes, investment levels, average income indicators, poverty and mortality rates, educational, health, and accommodation standards, as well as infrastructural estimates.[86] Presidential Decree No. 825 from 28 June 2007 'On estimating activity of the executive branches of power of the subjects of the Russian Federation' instructed the governors to forward annual reports on these parameters from September 2007. Under this system, the centre would find it extremely difficult to dismiss a politically inconvenient governor, who produced undisputable assessment results.

The assessment process has also remained highly bureaucratised, which could facilitate the production of enhanced indicators by administrative methods. Indeed, administrative institutions that are directly or indirectly controlled by the governors remained responsible for assigning nearly all these estimates. This left very little voice to civil society or population, and made drawing the objective picture of gubernatorial effectiveness difficult. It is also significant that the Kremlin has omitted a range of important political analysis indicators. These are directly connected with institutional performance and include such indicators as bureaucratic responsiveness, availability of statistical and information services, budget promptness, and the level of professionalism within executive and legislative institutions (see Putnam 1993, pp. 66–73; Stoner-Weiss 1997, pp. 90–129). Only four out of the existing 43 indicators measure governmental responsiveness: the level of satisfaction with regional medical service, the ability of the regional government to tackle crime, the citizen's satisfaction with activities of the regional leadership and their information openness, and the level of satisfaction with educational standards.

Moreover, from the point of view of policy analysis advanced by Putnam (1993), these indicators in many cases reflect the policy 'outcomes', not the policy 'outputs'. For example, some indexes include the number of people absent from work due to sick leave, or the number of people actively engaged in sports activities, or the number of people taking part in regional cultural programmes, or direct mortality and industrial accidents rates. Such outcomes are affected by a large number of extraneous variants. Mortality rates, for instance, are influenced by diets and life styles that are beyond the control of a governor; a rise in sick leave rates can be due to weather conditions; company profits might be subject to industrial discipline, motivation, and the country's economic performance that are beyond regional governmental control. From this point of view, the objective assessment of gubernatorial performance remains ever more difficult, which gives the regional leaders an opportunity to manipulate data, provide extraneous explanations for bad results, and negotiate these themes with the federal centre. In Chapters 5 and 6 of this book I will return to the problems of gubernatorial appointments and discuss further decentralising effects of this initiative in detail.

The right of the federal centre to initiate criminal prosecution against the governors has proven to be a powerful but largely decorative instrument. There were 11 criminal cases brought against Russia's governors between May 2000 and December 2004, i.e. during the period when the governors were elected officials.[87] Interestingly, none of these cases was concluded with a conviction and all charges were dropped following a brief period of investigation. It is very likely that the prosecution employed these powers with the sole purpose of tarnishing the image of regional leaders in the public domain prior to the election. Indeed, in most cases these developments took place before the launch of an electoral race and, on a number of occasions, such a strategy was successful – an example being the electoral failure of governor Vladimir Platov in Tver in December 2003.

When gubernatorial elections ceased to be a serious political problem, the investigation picture was largely repeated albeit the number of criminal cases brought up against the governors has been substantially reduced. There were no such cases in 2005, even though a Moscow Court decision placed the property of the Ryazan governor Georgy Shpak under arrest. In 2006, only two cases were initiated against the governors of Nenetsk Autonomous Okrug and Republic of Khakassiya Alexei Barinov and Alexei Lebed respectively, and one such case took place in 2007 against the governor of the Amur region Leonid Korotkov.

It is important to mention that in most such cases, the accusations related to previous allegations that have been long known to the legal enforcement agencies. Therefore, with the exception of Alexei Barinov, who received a suspended conditional sentence, none of these cases were finalised with a conviction.

The focus of attention, on the other hand, has been shifted to local mayors, as the remaining segment of the regional elected power. By the end of 2007, 17 mayors and their deputies had been arrested and prosecuted. In many such cases, criminal charges were related to genuine cases of serious mismanagement and embezzlement. This is particularly true, given that the mayors control the most lucrative aspects of industrial, investment, and construction activities that are traditionally permeated with crime and corruption. At the same time, a number of cases had politically driven motivations and were closely related to the conflicts between the regional and municipal levels of power. One such example concerns the mayor of Stavropol Dmitry Kuzmin. Kuzmin was charged with asset stripping and tax evasion mostly as a leader of the regional *Spravedlivaya Rossiya* SR (Just Russia) party branch, which managed to defeat *Edinaya Rossiya* led by the ex-governor Alexandr Chernogorov during the March 2007 regional parliamentary election (Sedlak 2007). Similarly, Putin publicly admitted that criminal charges brought against the mayor of Arkhangelsk Aleksandr Donskoi were related to his political conflict with governor Nikolai Kiselev.[88] In such cases, the federal government sided with the regional authorities, thus allowing the latter to resolve their political problems via explicitly coercive administrative methods.

I will now turn to the legal problems related to the division of responsibilities between the regions and the centre. The governors consistently demanded the expansion of their responsibilities in their press interviews, State Council hearings, and personal meetings with the president. These demands were met during the Kaliningrad State Council in July 2005. The president announced the Kremlin's decision to transfer some 114 responsibilities back to the regional level.[89] The governors regained the rights to execute policies in the spheres of urban development, forestry, land management, the environment, and cultural heritage (Stanovaya 2005). The list had been expanded by 15 additional responsibilities by December 2005. More importantly, the centre attempted to work in collaboration with the regions on the implementation of the responsibilities transfer programme. In September 2005 the Kremlin implemented the idea of the 'extended government'. In this system, the regional leaders take part in the meetings of the federal cabinet, give account of their progress within the regional socioeconomic spheres, and share their concerns on problematic developments.[90] It is important to mention, at the same time, that despite numerous requests and complaints, the governors were unable to secure the return of the 'two keys' licensing principle.[91] By the same token, licensing in the areas of gambling, access to natural water resources, and medicine remained under the federal jurisdiction, regardless of regional pressures.

Finally, the set of laws aimed at the pluralisation of political actors in the regions was mitigated by the June 2007 amendments. The new documents enabled the governors to dissolve regional and municipal legislatures and thereby exert substantial influence on the functioning of these institutions. The heads of regional administrations obtained the right to dissolve a regional or municipal legislature if the latter did not convene for three months for political reasons. The law also stipulated that the first legislative hearing must take place no later than 30 days following the election. It has also been established that the hearings could only be deemed legitimate if over two thirds of the incumbent deputies were present.[92] Such legislation enables the governors, who control majorities on their legislatures' floors, to orchestrate early elections and thereby exert a large influence on legislative process within their respective parliaments.

The influence of *polpredy*

The institution of plenipotentiary representatives has failed to act as a serious impediment to gubernatorial authority in the regions. As Kryshtanovskaya (2002, p. 164) puts it, 'the function of *polpredy* was not in real and systematic control over the governors but in providing the president with support when and where he needs it'. Orttung (2004, p. 23), at the same time, refers to the presidential envoys as 'an extra layer of state bureaucracy with very poorly defined powers'. Indeed, a comparison of powers exercised by the governors and plenipotentiary representatives does not speak in favour of *polpredy*.

First, during the first term of Putin's presidency the governors remained elected officials who drew their legitimacy from popular support, and this was an important power factor unavailable to plenipotentiary representatives. With the enactment of the gubernatorial appointment system, the plenipotentiary representatives have become responsible for selecting the candidates for presidential approval. However, this transpired to be a decorative power, as *polpredy* often learnt of the presidential nominees during the late stages of the appointment process. Moreover, the governors often negotiated their positions directly with the federal centre without consulting the plenipotentiary representatives. Thus, the governor of Adygeya Khazret Sovmen resolved his 2006 appointment problem independently from the then most influential *polpred* Kozak. Furthermore, Russian experts on regional relations insisted that governors remain regional figureheads and main political forces defining major vectors of socioeconomic and political development for public and private sector elites.[93] Polls conducted in autumn 2001 with 250 senior officials of regional administrations and governments at the Sociological Research Centre of the Russian State Service Academy support this view. All respondents suggested that regional power was concentrated mainly in the hands of the governors, while the presidential representatives occupied much lower positions in the list.[94]

Second, the governors retained the freedom to appoint their cabinets. In direct contrast to this, presidential envoys did not have a choice in the composition of their offices as the latter represented a result of nominations by the presidential administration. In accordance with presidential decree No. 849, *polpredy* were formal members of the presidential administration and the head of this institution signed the nominations of all *polpredy*'s deputies.[95] Unsurprisingly, only individuals personally known to Putin occupied these positions. In 2000, the list of proposed candidates for the post of deputy *polpred* in the North-West federal district comprised Vladimir Vyunov – a close friend of *polpred* Cherkesov (who is himself personally close to Putin), Alexandr Kuznetsov, a judo companion of Putin's during his St. Petersburg days, and other former colleagues or associates.[96] Kryshtanovskaya (2002, p. 163) notes that in 2002, 70 per cent of *polpredy*'s deputies and 34.5 per cent of chief federal inspectors were representatives of legal enforcement agencies, where Putin has much weight. This situation remained largely unaltered throughout the entire period of Putin's presidency. Two fifths of the subsequent cadre rotation was also drawn from the military structures and by 2006, the share of military men had grown to comprise one third of the entire *polpredy*'s office personnel (Petrov 2006, p. 83).

It is also significant that the *polpredy*'s positions were filled with candidates personally known to Putin. Most plenipotentiary representatives were either Putin's close colleagues and trusted friends or those who worked under his leadership in various institutions and were unlikely to make any sudden political moves. For example, both *polpredy* Poltavchenko and Cherkesov have been Putin's close friends and associates since his days in the Soviet secret service. *Polpredy* Latyshev and Drachevsky worked under Putin in the St Petersburg Mayor Office and in the Russian Federation Government respectively. Polpred Anatoly Kvashnin served as the Chief of Russia's General Staff and was thus subordinate to Putin. *Polpred* Yakovlev was Putin's defeated opponent from St Petersburg, and both *polpredy* Matvienko and Klebanov were members of Putin's St Petersburg cadre reserve.

Third, it is important to mention that the administrative resources of plenipotentiary representatives were unimpressive in comparison with those of the governors. The staff in federal districts' headquarters numbered on average 150 including chief federal and federal inspectors (Kryshtanovskaya 2002, p. 163), which is small given that the office is responsible for at least ten regions with much larger administrations. The 'administrative resource' of the presidential representatives was also marred by the fact that the centre, while instituting undoubtedly stronger plenipotentiary representatives, reappointed old personnel to these structures. According to a 2001 estimate made by the Moscow Carnegie Centre (see Zubarevich *et al.* 2001, p. 174), approximately a third of the 'old' representatives found employment in the new structures. For example, Georgii Poltavchenko, a plenipotentiary representative in the Central District, occupied the post of presidential representative in Leningrad Oblast; his new deputy Alexandr Bespalov was a presidential representative in St. Petersburg. The former presidential representative in Stavropol Krai Alexandr Korobeinikov became a deputy plenipotentiary representative in the Southern Federal District. Twenty former presidential representatives were appointed as head federal inspectors in the regions, mainly in the Siberian and Central Federal Districts during the early reform stages. Because former presidential representatives were loyal to the governors due to their heavy dependence on gubernatorial subsidies, it seemed practical to the centre to adopt such an approach to appointments in order to ensure a smooth institutional transition and not to aggravate the governors at a time of important change. This is not to say, of course, that many of these appointments evoked a positive gubernatorial reaction. The mayor of Moscow and representatives of the city administration, for example, complained that the president appointed Poltavchenko without any prior agreements and discussions with the city administration, which could impede the quality of their future co-operation.[97] The line towards appointing the 'old reserves' candidates faded after 2003, leaving just one sixth of local cadres in place. However, following the 2004–5 appointments wave, the number of regional staff in the *polpred* offices has grown again, thus leaving the situation substantially unaltered. Petrov (2006, p. 83) calculates that by 2006, the share of local appointees in the offices of plenipotentiary representatives had comprised as much as one quarter of the entire personnel.

Fourth, in the sphere of the relations between *polpredy* and the centre, Putin created a sophisticated institutional balance of power, which prevented presidential envoys from gaining excessive strength and implementing their own programmes independently of the federal centre. This represented a major difference between *polpredy* and the governors, who could legally and practically pursue their individual policy ends. One of the most important components of the administrative restrictions imposed on *polpredy* originated from the fact that the plenipotentiary representatives could not occupy their positions after the expiration of the incumbent president's term of office. This effectively implied a great degree of inter-dependence between central and federal districts cadres and precluded the envoy's rationale for expanding autonomy from the centre. Moreover, Putin's decree No. 849 required the envoys to pursue policies based on the federal constitution, federal laws, presidential decrees, and decisions made by the head of the presidential administration.[98] In this connection, Petrov (2006) claims that *polpredy* do not perform any independent functions and mainly implemented the policy of 'control' and promotion of the federal centre's interests in the regions. This is associated with the subordinate status of the presidential representatives and their close involvement with federal institutions – the president, the presidential administration, various federal Ministries, the Accounts Chamber, the Security Council and others.

Fifth, it is also important that *polpredy*, in contrast to the governors, have never been public figures, which also impeded their ability to pursue any independent policy line. In this connection, many of the former and incumbent *polpredy* who occupied the office between 2000 and 2008 have been representatives of military structures. Among them were general Georgii Poltavchenko of the Central Federal District, general Petr Latyshev of the Ural Federal District, generals Viktor Kazantsev and Vladimir Ustinov of the Southern Federal District, generals Konstantin Pulikovsky and Oleg Safonov of the Far Eastern Federal District, general Anatoly Kvashnin of the Siberian Federal District, general Viktor Cherkesov of the North-Western Federal District, and general Alexandr Konovalov of the Volga Federal District. For some of these generals, the appointment to the *polpred* position represented a serious career demotion and a consolation prize, following a dismissal from a more serious political position at the federal centre. Similar demotions took place among the non-military, civil *polpredy*. The former vice Prime Minister Valentina Matvienko was appointed as the plenipotentiary representative to the North-Western Federal District in spring 2003 before becoming the governor of the St Petersburg Federal City. The former St Petersburg governor Vladimir Yakovlev was demoted to manage the Southern Federal District during the same time. Similarly, the former vice Prime Minister Ilya Klebanov replaced Matvienko in the post of the North-Western District *polpred* in autumn 2003.

Moreover, the initial public activity of the plenipotentiary representatives has faded rapidly. Polpred Latyshev had to withdraw from any media activity, following his long-term conflict with the governor of Sverdlovsk region Rossel. Rossel, on the other hand, has had his media relations unchanged and almost figured as a victorious side in the dispute. Similarly, while the former Volga District *polpred* Sergei Kirienko declared his intentions to fiercely fight the most defiant governors of his region, the newly appointed *polpred* Alexandr Konovalov (2005) remained silent on all such matters.[99]

Sixth, legislative power also remained in the hands of regional elites, i.e. the governors and legislatures: in many regions such as Moskva, Voronezh Oblast and Moscow City, the governors initiate 70 per cent of the adopted legislation.[100] The exception is that this power now excludes the possibility of violating the federal code and facilitates the establishment of a 'single legal space' in Russia. Effectively, the governors lost the power to draft laws at their discretion if this violates the acting constitution. Plenipotentiary representatives implement permanent 'legislative monitoring', regularly attending sessions of regional assemblies and establishing permanent working commissions that check newly proposed regional legislation for compliance with federal standards.[101]

Finally, the economic resources of governors were incomparably greater than those of plenipotentiary representatives. Turovsky (2001) notes that all economic powers available to the regions were concentrated in gubernatorial hands and were unavailable to *polpredy*. Gubernatorial participation in the handling of important economic projects within federal districts is also an important factor that indicates the significance of the institution of the governors vis-à-vis plenipotentiary representatives. For example, the complex process of the regional merger between the Irkutsk region and Ust-Orda Buryat Autonomous Okrug took place with an active participation of the regional governors and with the seemingly visible absence of *polpred* Kvashnin.[102]

Equally important remains the fact that penetration of trans-national corporations into regional politics has changed the dynamics of socio-economic processes within the regions. Given that Russia's leading enterprises have emerged as significant regional actors, sub-national authorities and entrepreneurs have begun to look for potential forms of interaction. Big business is interested in the provision of a favourable legal and political climate for investment, while regional leaders are in need of substantial financial resources to implement various social programmes (Turovsky 2002). This process has left little room for the plenipotentiary representatives, whose role in the *business-governors* duo has become rather insignificant.

In conclusion, we can see that the process of federal relations in Russia remained complex and intricate, regardless of the centralising structural drive by the Kremlin. This created a certain inconsistency between the newly emerged hierarchical institutions and decentralising functional processes taking place within them. More importantly, the nature of such decentralising processes was consistent with Russia's unique political style. The regional drive towards autonomy has become adaptive, informal, non-transparent, and in many cases bordered on political conformism. The central policy of selectiveness and favouritism added its voice to this stylistic picture. Thus, the balance of power between the two orders of government has changed its structural institutional expression but retained its fundamental personalised, informal and decentralist nature. In the following chapter we will move on to the discussion of the supporting pillars of federal institutions and the ways in which political actors implemented and operated them in Russia.

4 Institutions of centre-regional integration

Monocentrism and its potential implications

The monocentric power system

In order to analyse the dynamics of federal integration in Russia we will employ methodological suggestions of analysts from the Moscow Centre for Political Technology (Zudin 2002). At the level of functional power arrangements, these scientists suggest that political processes taking place within Putin's institutional systems of federal integration should be referred to as monocentric and contrasted in principle with Yeltsin's style of polycentrism. The principal difference between polycentric and monocentric functional logics is manifest in the level of authority of the federal centre within the existing political framework. Polycentrism assumes multiple centres of influence – the president, oligarchs, regional governors, local mayors, and the national legislature. Within such conditions, Yeltsin's presidency always relied on a close alliance between the Kremlin and various segments of the existing elites. A monocentric operational pattern, on the other hand, has a sole power centre and bases its stability on an unofficial 'social contract' between the centre and society, often bypassing the consent of the elites. Therefore, monocentrism assumes central authority as the cornerstone of the existing political framework, while other important elements of the political system – financial elites, regional leaders, legislators – act not as alternative centres of political power but as functional components of the decision-making process placed at a certain distance from the centre. The phenomenon of 'Putin's electoral majority' could potentially be regarded as a provision for a monocentric 'social contract' and a possible basis for a monocentric operational arrangement. In this context, the Kremlin's break from the overbearing pressure of old power echelons that took place under Putin's leadership could be viewed as a factor which enabled a wide-scale reform of Yeltsin's functional power system.

This chapter will argue that Putin concentrated on subordinating all alternative power centres to a single authority core in the Kremlin. In pursuit of these goals, the president came very close to establishing an unequivocally monocentric power redistribution and secured a certain stability in institutional interrelations. This was accomplished by various means: consolidating presidential/central authority, distributing authority among institutions in a power dependent style, altering the composition of the political elite, striking limited deals with certain segments of the elites, and confining the conflicts within given institutions, thereby preventing a spill-over into the public sphere. The resurfacing of these types of interactions is consistent with the reactive influence of Russia's political style, which is based, as discussed, on informal power arrangements, personification of authority, and arbitrary political conduct.

Another important aspect of this discussion concerns the underlying differences between a monocentric functional system and integrated co-operative arrangements practised by the world's most successful federations. The discussion treats mechanisms for effective centre-regional co-operation, various degrees of federal integration, and factors that ensure the socio-political stability of successful federal countries. A number of academics (Ordeshook and Shvetsova 1997; Elazar 1997; Duchacek 1970; Duchacek 1986; Burgess 1986) often point out that Western models of federal democracies have much stronger levels of inbuilt stability and greater institutional effectiveness because the regional political elites in these countries rely on a well-established system of institutional incentives to co-operate with the federal centre on a formal legal basis. Such a system requires certain commitments by the public and politicians to the values and culture of federalism. As outlined earlier, the most significant aspects of such values include the ideas of tolerance, reciprocity and respect towards individual integrities of constituent territorial entities united into a single whole on a contractual constitutional basis (see Davies 1978, p. 3; Elazar 1997, p. 33 and p. 67; Duchacek 1970, p. 192). In this context, partnership and recognition, as opposed to hierarchy and domination, constitute the central tenets of federal ideology and operational culture of genuinely federative institutions. And the commitments of political actors to these values represent, perhaps, the most important facet of federal political systems. Monocentrism, in contrast, largely relies on the authority of the federal centre and often employs coercive and administrative powers to 'persuade' the regions to co-operate with central institutions.

At the structural level, in this chapter we will demonstrate that, in some key policy areas, the president and his team of reformers aimed to pursue the modernisation of the regional institutional system following a blueprint primarily based on Western examples. Indeed, a number of Putin's early policies moved Russia away from a semi-confederal, treaty-constitutional arrangement and brought the country into greater conformity with the institutional requirements set out for federations. In the previous chapter we have already discussed the government's efforts at ensuring the supremacy of the federal constitution and an adequate division of powers. In this chapter we will analyse the emergence of the federation's supporting structures. I will argue that the federal centre made further attempts at promoting multilateral intergovernmental co-operation in the three most important directions: (1) erection of effective structures of intergovernmental dialogue; (2) establishing the system of integrated national parties; and (3) promoting centre-regional migration of cadres. However, all attempts made during Putin's presidency to reform the regional system were implemented in Russia's traditional political style, i.e. using a 'top down' method, which predetermined the major distinction between Putin's monocentric federal design and Western models of successful federal states. The overt reliance of the functional parameters of Russia's federal system on the arbitrary role of the president provided the basis for creation of a virtually monocentric institutional arrangement, in which all alternative poles of political influence are practically transformed into central political satellites. This ensured that the new regional model remained unco-operative, non-integrated, and adaptive in form. In this context, the functioning logic, the culture, and the ethics of Russia's federal institutions represent the most important political problem. These cultural and ethical limitations do not deliver an optimistic message for the potential direction in which Russian federalism is headed. Hence the ability of the federation to survive various environmental challenges, such as economic downfalls or a change in the Kremlin's political leadership, is under question. In this chapter we will first discuss the emergence of Western-based structures and then move on to an analysis of the functioning logic of these networks.

Emergent structures and initial achievements

Intergovernmental networks

Russian political scientists have been examining the importance of interdependence between the functioning of central and regional institutions and had pointed out the importance of interconnection and permanent inter-exchange of information between regional and central layers of government. For example, Alexandr Mikhailov (2001, p. 13), a sociologist at Moscow State University, insists that a national institutional structure represents a self-contained system in which adequate information channels ensure certain stability in the functioning of this system. These channels can be established through the formation of a range of co-operative institutions in functional areas of economics and politics. These institutions are necessary to gauge the reaction of sub-national political players to policy strategies adopted/proposed by the actors at national level. According to Mikhailov, such an arrangement would ensure constant inter-institutional dialogue between differing echelons of power and establish a system of balances preventing both sides from going beyond or below their limits of authority.

The issue of limits of authority is particularly salient: if the centre or the regions go above or below certain established thresholds of power, the entire system could become unbalanced and in extreme cases would head either towards authoritarianism or towards anarchy. This view was practically supported by Badovskii (2001). He insists that the parameters of the system of central control over the regions seen under Yeltsin fell substantially below the 'manageability threshold', which resulted in a factual disintegration of the existing federation, legal and political anarchy, and the creation of a virtually confederal state structure. Putin's administration understood the significance of intergovernmental structures for Russia's federal dynamics. Intergovernmental networks emerged at various levels: presidential, legislative, governmental, and in the Federal Districts that were instituted in May 2000 as an additional administrative level of centre-regional dialogue.

The State Council was established on 1 September 2000 to promote consultative links between the Russian Federation president and the governors. The Council convenes four times a year and its presidium meets prior to these events. This body functions with the aid of 22 working groups. Each group is responsible for specific developmental areas such as transportation, social policy, ecology, international relations, local government, land reform, and taxation, and comprises representatives of regional and federal officials at the highest ministerial and gubernatorial level.[1]

The Legislative Council was instituted on 21 May 2002 as an inter-parliamentary consultative structure. The Council meets twice a year and its presidium, which includes the heads of selected regional assemblies, the leadership of the upper chamber of the Russian parliament, and seven plenipotentiary representatives, convenes at least four times per annum. The president and several ministers normally take part in the Council hearings, which also include representatives from the Federal Assembly, regional executive branches, local self-government, and non-governmental organisations. This institution has eight working committees, which are chaired by heads of sub-national legislative bodies on a rotating basis.[2]

The Federation Council – the upper chamber of the Russian parliament that was previously composed of the heads of regional executives and legislatures – has been restructured to comprise one appointed representative from the executive and legislative branch of each region.

Federal programmes, which represent developmental initiatives aimed at levelling regional disparities, have become an avenue of collaboration between the central and regional governments. Fifty-three such initiatives were registered for implementation in 2005 across six different subdivisions accommodating policies vis-à-vis the development of infrastructure, education, judicial reform, ecology, economics, and the regions. Moreover, thirteen permanent co-ordinating committees and councils were established under the aegis of the federal government to promote integration in socio-economic, political, and cultural matters.[3] The creation of seven Federal Districts, headed by presidential envoys, was intended tp expand centre-regional institutional co-operation, through which regional political elites could have a quick and reliable access to the federal centre.

To some extent, these institutions created new opportunities for collaboration between the centre and the regions and enabled the regional elites to become involved in the process of state policy formulation. The State Council has become one such structure. For example, the governors collectively submitted proposals to the State Council concerning innovations in the sphere of small and medium business taxation, which the president transformed into a legislation project and introduced to the lower house of parliament in spring 2002.[4] Similarly, in November 2002 the State Council mitigated the effect of proposed presidential amendments to the law on 'organisational principles of legislative and executive bodies of the subjects of the Russian Federation'.[5] As discussed in Chapter 3, Putin's original proposal was far too constrictive providing for a number of administrative shortcuts for gubernatorial dismissal. The final edition of this document, adopted by the Federal Assembly in June 2003, retained the main provisions of the Constitutional Court decision from 4 July 2002 that established a lengthy procedure for the removal of governors from office.[6] The reform of local government (Lankina 2004) was perhaps the Council's most important legislative engagement of 2002, as this level of authority traditionally represents one means of undermining the power of regional governors from below. The new initiative, drafted by the Kozak committee, provisioned greater control over the activities of mayors in large cities and introduced a clearer delineation of financial flows to various types of localities. In July 2005, the State Council meeting adopted a range of important measures that expanded the rights and responsibilities of the regional governors. This marked the beginning of a new phase of the regional reform. The 19 February 2007 State Council meeting pondered such strategic issues as the adoption of measures aimed at the development of Russia's industrial sector and modernisation of the national economy. Similarly, the Legislative Council took an active part in the discussion on the division of responsibilities between the central and regional levels of power, reform of the regional electoral system, and local self-governance, as well as the land and forestry regulations.[7] With regard to the State Council, it is important that the governors regularly attend this body's hearings with the view of delivering their opinions on various developmental matters. Moreover, governmental ministers, representatives of big business and academia, members of both houses of parliament and the presidential administration often take part in the Council hearings as to make the dialogue more functional and inclusive. Given that the Council have often been involved in the resolution of strategically important problems, on 23 February 2007, Putin issued decree No 241 that expanded the Council membership to include ex-governors of Russia's regions. This measure aimed at enhancing the centre-regional dialogue and making the discussions on pressing political issues more objective.

Finally, from a theoretical point of view, the new method of the Federation Council formation provided the governors with a room for manoeuvre in appointing the senators. The law 'On the principle of formation of the Federation Council of the Federal Assembly of the Russian Federation' provides the governors with the right to nominate the senator from the region's executive branch while the regional legislatures will have to approve the nomination in a vote.[8] For regional legislatures, it will require a two-thirds majority vote to block the gubernatorial nomination, which practically guarantees its success. Therefore, the law ensures that governors would have a strong chance of assigning the senators of their liking, who would then be personally indebted to them through these appointments. With respect to the representative branch, the law ensures that either the head of the regional legislature or a group of deputies comprising more than one third of the regional Assembly could nominate the candidate. However, the poor level of development of the party system makes it highly doubtful that such a large group of deputies can emerge to put forward a nominee. Moreover, given that the governors normally dominate the assemblies (and respectively their leaders) the law gives them a chance to informally approve both nominations.

Parties

Attention has also been paid to the creation of a national party system. However, during the first two years of Putin's presidency, the regionalisation of national party organisations created serious political difficulties for the centre. Indeed, by 1999 the governors realised that from a tactical point of view, local parties can turn into a very powerful resource of influence not only on regional but also on national issues. Starting from this period, the regional leaders began to take control over single mandate districts in order to form with candidates from these areas strong pro-gubernatorial coalitions within regional legislatures. Very often, regional governors created local parties whose representatives ran for election to representative institutions. Moscow Carnegie Centre analyst Dmitrii Olshanskii (2000, pp. 45–6) has observed that most regional parties and 'spontaneously' created factions of regional legislations were 100 per cent dependent on governors in both the financial and administrative spheres. It is not surprising that these parties often changed their opinions and platforms in accordance with the current interests of the governors. It is also interesting that national and regional parties had the tendency to permit mixed memberships: it was often the case that participation in a national faction coincided with that in a regional faction. More importantly, these multi-faction memberships did not follow a particular ideological pattern in which all members of one regional faction have an association with a particular national faction.[9] This situation indicates that regional and national parties had no ideological/political affiliation – a fact that resulted in the 'drain of cadres' from national to regional party structures in accordance with current political-administrative advantage. In addition, it was often the case that those regions which were 'problematic' for the Kremlin had strong regional parties in regional legislatures. For example, in Sverdlovsk Oblast – an initial bastion of opposition to Putin's federal reforms – regional parties managed to provide a strong alternative to the national parties. During the April 2002 elections, the regional electoral block *Za Rodnoi Ural* (*For Homeland Ural*) attracted 29.45 per cent of votes corresponding to seven seats. Federal parties gained support that was much more modest: *Edinaya Rossiya* with 18.36 per cent and the CPRF with 7.3 per cent and the *Party of Pensioners* with 6.11 per cent of votes.[10] In St. Petersburg, which during the first three years of Putin's presidency has had a very controversial relationship with the centre, the legislature comprised four regional factions counterbalanced by four national factions.[11]

In this context, Putin's administration realised that the system of federal parties would enable it to solidify its political autonomy from regional elites, reduce the influence of local economic clans, and politically integrate the current federal system (Orttung and Reddaway 2004, pp. 41–2). The Kremlin's plans for the development of political parties were disclosed by the deputy head of the presidential administration Vladislav Surkov at a meeting with the leaders of *Edinaya Rossiya* in February 2002. In his speech, Surkov pledged that participation in national elections would be transformed into a prerogative of national parties – a policy that would seriously diminish the positions of regional bogus parties serving the private interests of local business groups and the governors. Surkov also confirmed the Kremlin's intention to transform national parties into an integral part of the state institutional system and enable them to take an 'active part in regional gubernatorial elections and in the process of formation of the national government and other executive institutions of national importance'. Finally, he declared that all high officials on the level of ministers, governors, and presidents of the republics would receive the right to participate in the functioning of political parties by 2004.[12] Bunin *et al.* (2002) suggest that the Kremlin's party-building project included three important elements: (1) the establishment of the Kremlin's own party of power *Edinaya Rossiya* and satellite projects such as *Rodina, Spravedlivaya Rossiya, Grazhdanskaya Sila*, and the Democratic Party of Russia, (2) increasing the role of the loyal State Duma in order to reduce the influence of bureaucratic and oligarchic clans in the federal government during the first term of Putin's presidency, and (3) the issue of legislation conducive to the development of national parties.

First, the Kremlin has decided to assume the leading role in the party building project and to take control over the developments evolving in all wings of Russia's political spectrum. *Edinaya Rossiya* (ER), as the dominant ruling party of Russia, represents one of the main achievements of the presidential administration. Putin has also attempted to take control over the left-wing partisan dynamics. The left-patriotic block *Rodina* represented another political creation of the Kremlin, which was initially designed to split the Communist electorate. However, while the block successfully implemented the Kremlin's task during the December 2003 parliamentary election, it has subsequently failed to become a viable political force with a solid independent programme. Soon after the election, the block disintegrated due to internal disagreements over opposing Putin in the March 2004 presidential race. By 2007, the federal centre had decided to take control over the electorate, which was dissatisfied with the ER performance, viewed it as an overtly administrative organisation, and voted for the existing socialist projects such as the Party of Pensioners, Party of Life, *Rodina, Yabloko* and other smaller movements. The size of this electoral slice was estimated to be around 20 per cent of the entire voting market (Makarkin 2007). Thus, the *Spravedlivaya Rossiya* (SR) party project was launched in spring 2007 to implement this task, and it is within this light that the party was often referred to as the 'second pillar of the presidential administration'. The party, led by the Kremlin's loyal Federation Council speaker Sergei Mironov, was composed of the leading regional actors *Rodina*, the Party of Pensioners, and the Russian Party of Life. Attempts at controlling the right wing of Russia's opposition spectrum were practically visible with the support of the *Grazhdanskaya Sila* (Civic Force) and Democratic Party of Russia projects during the December 2007 and March 2008 electoral campaigns. These dynamics reached their apex in September 2008 with the dissolution of the Union of Right Forces (SPS) and the building of the pro-Kremlin right wing project comprising the remains of the Union of Right Forces, the Civic Force, and the Democratic Party of Russia.[13]

Despite a large number of limitations, the Kremlin's drive towards establishing its personal 'parties of power' has led to some unexpected positive developments. The December 2003 achievements of *Edinaya Rossiya* and *Rodina* were rather impressive. *Rodina* gained as much as 9 per cent of the vote, and nearly 18 per cent of all registered electors cast their vote for the *Edinaya Rossiya* party. This corresponded to 37 per cent of all those who turned out to the ballots.[14] The December 2007 situation improved the ER positions even further. It is important that the turnout surpassed its previous 2003 levels, amounting to some 60 per cent of the population. *Edinaya Rossiya* almost doubled its result collecting some 64.1 per cent of the vote.[15] Furthermore, consistent attempts to build a controllable opposition on the left of Russia's political spectrum eventually resulted in the acceleration of party struggles in the regions. Indeed, despite the monolithic partisan façade at the national level, the regions displayed some serious signs of political diversity. While *Rodina* showed positive developmental dynamics on the regional landscape during the entire time of its existence, this trend accelerated with the launch of the SR project. In many regions, such as Stavropol Krai, Samara, Vologda, Leningrad, and Tomsk Oblasts, SR managed to consolidate local political elites against the ER party. Many ER-SR conflicts – such as those in Stavropol, Samara, and St Petersburg[16] – spilled into the national arena. These dynamics seriously questioned the de-facto dominance of a single pro-Kremlin party, prompted the existing political organisations to refine their ideologies (Zvonarev 2007), and created conditions for effective coalition building within the regional legislatures (Ryabov 2007).

Second, the functioning of the third State Duma markedly differed from its predecessors, which at the time created some grounds for cautious optimism with regard to the party formation process. Arguably, the third Duma became a much more stable political institution, in which the president could rely on a parliamentary majority capable of conducting a constructive dialogue with the government. At the same time the opposition, mainly supported by the business elite, also had a solid political potential in the legislature. This group exerted its influence via the establishment of a number of official structures such as an inter-faction association 'Russia's energy' headed by the representative of *TNK* Vladimir Katrenko, and via the informal lobbying groups mainly co-ordinated by the representative of *Yukos* Vladimir Dubov (Makarkin 2004). These groups helped to block a number of the Kremlin's initiatives aimed at curbing the interests of the large oil producers – an example being Kozak's idea to replace mining licences with concession agreements. Moreover, the existence of internal off-shores, which benefited Russia's oil giants and had devastating consequences for economic development in many regions, represented an insurmountable fortress for the federal government. In 2002, the parliamentary oil lobby succeeded in blocking a series of laws that aimed to abolish regional tax benefits for investors. A similar situation occurred vis-à-vis the law 'On Customs Tariffs' that planned to enable the government to establish export duties on refined oil produce independently of those on unrefined oil.[17] This package was accepted only at the end of 2003, when Putin's attack on *Yukos* changed the parliamentary atmosphere and the attitude towards lobbying by oil producers.

In addition, comments by a number of opposition members that Putin managed to subordinate the third State Duma to his arbitrary institutional system should be accepted with some reservations. On a number of occasions, parliament demonstrated the ability to pursue its own institutional interests. The 2003 pre-electoral developments demonstrated that the third Duma was able to legislate independently of the Kremlin: at the end of 2002 and beginning of 2003, the lower house rejected a number of unpopular presidential proposals. These concerned such important policy issues as reform of communal services, reform of the Russian electricity monopoly *RAO UES*, and ultra-liberal currency reform, which had been inspired by Putin.[18]

The multi-dimensional disposition of forces in the third State Duma and the political struggle associated with it represented a marked difference from Yeltsin's binary arrangement, in which the ruling elite and the opposition monopolised two separate institutional poles – the government and the legislature – and the political struggle unfolded mainly in the Duma-government (Kremlin) direction. In these conditions the parliament, which represented the sole opposition to the government, adopted a multitude of laws that were impossible to implement in the existing economic climate. In the mind of Alexandr Zhukov, member of parliament of all four convocations and Russia's deputy Prime Minister, this situation downgraded the State Duma to an institution whose opinions were not to be taken seriously.[19] In these circumstances, the president was often forced to rule by decree against the wishes of the State Duma in order to ensure the implementation of some of his initiatives. It is clear that the third Duma had a differing mode of operation and became a stronger institution with the ambition of establishing a 'dominant' party and forming a future government with such a party.

Finally, the Kremlin has adopted an extensive network of mechanisms aimed at the development of national parties at the sub-national level. Putin introduced three laws concerning the management of political parties in the country's politico-institutional arena, nominating candidates from national parties for the presidential elections, and providing the main guarantees of electoral rights for populations' in the regions. All these documents were intended to create a system of incentives for national parties to take root in the regions.

The law 'on political parties', which established the essential legal basis for the development of a full-fledged multi-party system, was enacted on 11 July 2001. The law demanded that Russia's parties must include no less than 10,000 members, have established branches in more than 45 regions of the federation counting no less than 100 people in each branch, and register at the Federal Ministry of Justice.[20] (The membership requirement was increased up to 50,000 in late 2004, following the Beslan events.[21]) It is important that the law remained silent on the issue of regional parties, effectively creating an unfavourable legal framework for the development of 'bogus' regional formations.[22]

The law 'on the election of the president of the Russian Federation', adopted by the Federal Assembly on 27 December 2002, increased the role of parties in the political and electoral process and was considered no less important than its predecessor. The document allowed political parties to nominate their candidates for presidential elections without the prior collection of signatures of public support.[23] All parties that entered the national legislature fell under this legal act in 2004. In contrast to party nominated candidates, independent candidates must present two million signatures of provisional electoral support in order to obtain official registration for the presidential race. A similar regulation applied to those parties which did not pass the electoral threshold to the national assembly. Within the new system, regional and national party-affiliated politicians received equal opportunities to compete for the highest executive office. The former head of the national electoral committee Alexandr Veshnyakov evaluated the new system as very logical and suggested that it could encourage the development of political parties organised on the lines of national – not personal or financial – interests.[24]

Finally, the law 'on main guarantees of electoral rights and participation of the Russian Federation citizens in a referendum', adopted in May 2002, required that half of regional assemblies' deputies must be elected through party lists. Regional electoral committees have also become more accountable to the Central Electoral Committee, thus precluding regional financial and bureaucratic groupings from influencing elections. In particular, local electoral committees were stripped of the right to deny potential candidates electoral registration: the law transferred this prerogative to the Court. The list of reasons sufficient to deny a candidate electoral registration was also substantially reduced. Furthermore, the law provided that gubernatorial elections must take place in two rounds, even if only one candidate remains for the second voting cycle. The practice of preliminary voting was also abolished.[25] These innovations were aimed at diminishing the administrative resources of governors and local economic elites in the sphere of regional elections by forcing them to share their authority with federal-level administrative and political players such as the Central Electoral Commission and national political parties.

These developments prompted political parties to improve their internal administrative structures, which was necessary for registration at the Ministry of Justice, and to create strong judicial and methodological departments that enabled them to act more effectively at the federal and regional levels. Thus, in comparison to the December 1999 elections, not a single party or electoral block was removed by the Central Electoral Commission from the December 2003 campaign (Borisov 2004).

Moreover, the introduction of party lists for the elections to regional assemblies led a number of parties to adjust their electoral strategies and tactics to regional political realities. In April 2007, for example, the SPS former leader Nikita Belykh emphasised the success and importance of welfare state in West

European economies, thus attempting to explain the practical benefits of liberalism to the general public (Zvonarev 2007). A similar situation surfaced within the Kremlin dominated parties. While the ER has infamously declared that 'supporting the policies of president Putin'[26] encompasses its ideology, the SR has presented the public with a clear socialist stand, refused to deploy Putin's images in its electoral campaign, and declared its intentions to rename the movement into Russia's Socialist Party.[27]

These developments could send a positive message to the Russian party-building project. In the optimistic scenario, this dynamic has the potential for breaking the dominance of a single pro-Kremlin force and creating a greater political participation and pluralism in the regional partisan politics. Moreover, while drafting ideological positions, current parties inadvertently attempt a certain convergence with the electorate and could better reflect the existing societal cleavages and platforms. This could represent a move away from the current, as well as the Yeltsinite partisan models, in which the country's political parties ultimately reflected the interests of political and financial elites. We have outlined some major trends within the party-building at the wide national level. Chapter 6 will further elaborate on the regional dynamics, as well as their causes and consequences.

Migration of cadres

In terms of the centre-regional alteration of personnel Putin made sure that certain key figures in regional politics entered the central elite. For example, the Federation and State Councils represented institutions in which a number of regional politicians – such as former president of Yakutiya Mikhail Nikolaev, the former president of Ingushetiya Ruslan Aushev, the president of Tatarstan Shaimiev, and many others – were given the opportunity to express their political talents. At the same time, these developments did not follow a transparent institutional pattern and depended on the arbitrary choice of the centre and its strategic policy goals towards certain territories. These problems will be detailed in the following subsection.

Operation and culture: the commitment problem

Despite the emergence of these positive signals and the erection of many important institutional structures, the *way* in which such institutions were set up to function permitted excessive centralisation and institutional hierarchy. These networks have been mainly operated in an informal and arbitrary political style, thus tarnishing the original modernising intentions. Indeed, in the area of institutional interaction among agencies responsible for the development of centre-regional dialogue, the Kremlin established itself as a solid centre of influence which relied on a so-called *power-interlock* mechanism, achieved via the methods of appointment and legislation. A new working formula for those central institutions responsible for the management of centre-periphery relations operated on the principle 'no president, no power', which lay at the heart of the monocentric functional arrangement. In this system, institutions such as the State Council, the Security Council, presidential representatives, political parties, and the system of cadre management were power-interlocked through both joint decision-making and the cross-institutional appointment structure, while the president acted as a figurehead and an arbiter. The pursuit of such an arrangement required substantial restructuring of the ruling elite that became established under Yeltsin. This assumed an alteration of the existing informal rules as well as the institutionalisation of new formal regulations that could ensure greater resilience of the new political system. On the institutional front, many initiatives, though not directly contradicting the constitution, made the regional order of government institutionally subordinate to the federal level, thus breaking up the federal promise of reciprocity. On the level of centre-regional operational communication, existing co-operative institutions denied the regions effective representation and thus impeded the formal and transparent dialogue between the two governmental tiers. The resulting consolidation of the monocentric functional process has seriously tarnished the evolution of the genuine federal model in Russia and impeded the chances of achieving federal stability. Let us discuss the evolution of these dynamics within the three aforementioned institutional dimensions.

Intergovernmental networks

The functioning logic of the emerged network of intergovernmental co-operation has given cause for concern in that it has greatly overvalued the role of the federal centre. This primarily concerned the Kremlin's influence over the composition of political players participating in these organisations and the range of questions discussed.

First, the Kremlin has always shielded itself from all attempts by regional leaders to use co-operative institutions to alter the balance of power to their advantage. For example, the plan by Tatarstan's President Mintimer Shaimiev to establish a new institutional division of responsibilities between the centre and the regions did not even get discussed after being denied a hearing in the State Council (Bunin 2002). In addition, while allowing third and fourth terms for the governors in summer 2002, in November Putin initiated a range of legislative amendments allowing the suspension of regional administrative bodies if a region's debt to the centre exceeded 30 per cent of its budget or if the region mismanaged federal subsidies or failed to organise effective crisis management. The governors failed to mitigate the effects of these initiatives during a series of State Council hearings, and with some amendments, the law was enacted on 4 July 2003.[28] The reconstitution of gubernatorial elections, enacted in January 2005, further devalued any compromises reached in the gubernatorial cadre management sphere.

The introduction of the institution of plenipotentiary representatives has further increased central leverage over the evolution of inter-regional relations.

The carefully crafted institutional appointment policy discussed in Chapter 3 of this book enabled the president to use plenipotentiary representatives as loyal politicians in the regions with significant authority potential. Badovskii (2001) notes that the inclusion of *polpredy* into the presidential hierarchy established them as powerful regional figures and underscored the effectiveness of employing them as advocates of central interests in the regions. Direct access to the president provided plenipotentiary representatives with a potential to counteract regional officials successfully and ensured their major political victories on behalf of the centre. Regardless of the visibly inferior power potential of the *polpredy* vis-à-vis the regional governors, the enactment of this institution managed to assist the federal centre in selected policy areas. Plenipotentiary representatives established differing policy lines towards various regions; these policies at all times converged with the centre's intentions to pursue certain political goals in certain areas. In Ingushetiya during the April 2002 elections, for example, deputy *polpred* Zyazikov won presidential office with visible assistance from the office of the plenipotentiary representative – a development that illustrated the centre's ability to pursue successful policies in ethnic republics. In the Volga Federal District, former *polpred* Kirienko effectively responded to central concerns regarding the composition of the government in the republic of Tatarstan, which had changed little since the Soviet period. The *polpred*'s office directly influenced the rotation of cadres in the regional government and by 2002 the average age of a minister was between 35 and 37 years – approaching that of the *polpred*'s officials.[29] Moreover, because Tatarstan represented a strategically important republic for the centre, Kirienko made a personal contribution in opening up the region's business sphere to inter-regional contacts and employed his administrative powers in defending the interests of Tatarstan's producers.[30] On the other hand, the President of Bashkortostan Rakhimov was often a subject of criticism from Kirienko's team in connection with his Soviet-style governance and the involvement of his family in the region's political life. The *polpred's* officials often indicated that they were prepared to make administrative interventions in the course of the region's presidential elections in 2003, during which they expected Rakhimov to employ his position and influence the results of the poll.[31] Remarkably, these developments coincided with the centre's institutional dispute with Bashkortostan over the future of the republic's constitutional arrangement.

Second, co-operation within the intergovernmental institutions has extended only to selected policy areas and encompassed initiatives that could not seriously challenge the centre's fundamental vision of economic and political reforms. For example, the government ignored the State Council's recommendations on the strategic programme of economic development drafted by Minister of Economics German Gref.[32] Similarly, ex-Prime Minister Kasyanov blocked the Council's proposals for the reform of the Unified Energy System (*RAO UES*) electric monopoly in favour of Anatolii Chubais's plans, and torpedoed the Council's suggestions for education reform (Orttung and Reddaway 2004, pp. 29–30). The vast majority of the Legislative Council's projects that diverged from the Kremlin's main developmental strategies were also ignored. For example, regulations on the sale of arable land were adopted despite the receipt of negative references from 33 of Russia's regional parliaments.[33] Similarly, the government and the State Duma did not take into account the concerns of regional legislatures about Dmitrii Kozak's plans for the division of powers between the centre and the regions in February 2003 and disregarded the Council's note of protest on the new forestry regulations in September 2004.[34]

Third, the Kremlin has followed the logic of 'selective co-operation' with chosen actors and restricted the circle of political players involved in the meaningful dialogue with the federal authorities. Those who were in the league of the 'privileged' often have represented the regions or institutions whose support was crucial for implementing specific policy goals pursued by the federal government at the national level. For example, the biggest and most effective federal programmes were launched mainly for selected regions of 'strategic importance'. In particular, developmental programmes for the relatively rich republics of Tatarstan and Bashkortostan were adopted with the purpose of winning the political support of these territories while implementing the centralising reforms of May 2000. Such programmes practically ensured an influx of federal transfers to these regions' budgets numerically similar to their Yeltsin-era fiscal privileges and thus mitigated the introduction of a stricter fiscal discipline and the new redistribution of taxation between the regions and the centre.[35] Analogous comments could be made concerning the programmes devoted to the south of Russia (with a separate programme established for the Chechen Republic) and the Kaliningrad region. The latter represents an area of strategic importance for Russia's dialogue with the European Union, while the former is crucial for securing the country's geostrategic positions in the Caucasus, in particular for the relations with Ukraine, Georgia, and Azerbaijan. At the same time, with the exception of Chechnya, a large number of regions could be equally worthy candidates for such federal assistance and required economic harmonisation more than the existing grant receivers, in particular the republics of Tatarstan and Bashkortostan which fall within the league of Russia's most successful territories. In 2005, the Minister of Economics Gref reiterated the central state intentions to pursue the programmes of 'strategic importance' and pledged to terminate some 40 existing initiatives by 2006. The 'black list' included the Kuril Islands developmental programme, seven ecological, three transportation, 19 economic initiatives, and 12 programmes that were due to expire in 2005.[36]

Similarly, meaningful co-operation within the federal government's co-ordinating committees and councils was adopted only for matters of strategic importance. For example, to ensure a smooth implementation of the controversial reforms of social benefits, the former Prime Minister Mikhail Fradkov established special co-ordinating meetings with regional governors on a weekly basis.[37] In addition, the most influential and active committee was the Council on Entrepreneurship that comprised leading representatives of Russia's big business and yet again served a 'strategic' purpose for forging co-operative 'equidistant' relations between the Kremlin and Russia's large corporations – one of Putin's main objectives on taking office in March 2000. The Committee on Co-ordinating Intergovernmental Relations (instituted on 4 October 2005) was tasked with tightening the centre's grip over the regions by supervising regional cadre policy, international relations, and the implementation of economic programmes.[38]

Fourth, the Kremlin aimed to institutionalise co-operative relations with the existing elites on a consultative basis, without taking the process onto a distinctive political level. While collaborating with various elites, the Kremlin was unwilling to grant co-operative institutions legal functions and preferred to remain the ultimate arbiter of the political process. For example, the Council on Entrepreneurship created under the aegis of the government and the RSPP (Russian Union of Industrialists and Entrepreneurs) had purely consultative relations with the Kremlin. The Kremlin's predominance over the Union of Industrialists and Entrepreneurs was demonstrated during the *Yukos* affair, when the Union's members refused to take a tougher stance in defence of their corporate interests. In addition, the Kremlin attempted to develop a similar level of relations with the State Council, precluding all attempts to transform this institution into a venue for political bargaining between the centre and the governors by granting it constitutional status. Similarly, all ideas to introduce an electoral principle for the formation of the Federation Council remained on paper. Bunin (2002) observes that the Kremlin established a certain 'division of labour between the State and Federation Councils: the first institution was meant to act as a venue for reconciliation and agreement, whilst the second pursued purely legislative goals'.

Finally, the centre has begun to determine the scope and subject matter of discussions that take place in the Federation Council and the composition of actors that were allowed to take part in these processes. Putin's first step in this direction was lobbying for the appointment of Sergei Mironov – a former colleague from St. Petersburg – to head the upper chamber. Given that the political role of the Federation Council is directly interconnected with the political performance of its leader, Mironov's nomination ensured that this institution would be closely associated with the president and complemented the pro-presidential power coalition. Shortly after his appointment, Mironov brought a number of pro-Putin politicians from St. Petersburg to head important posts in the Council. The position of Mironov's media advisor went to Lyudmila Fomicheva, a former press secretary for the St. Petersburg mayoral office, while Petr Tkachenko, a former manager of St. Petersburg's mayoral administration, assumed responsibility for heading the Federation Council's apparatus.[39]

Moreover, the presidential administration played an important role in determining the choice of senators delegated to the upper chamber. We have already mentioned Remington's (2003, p. 674) conclusions that 75–80 per cent of the appointments were either recommended by or cleared with this body. The pressure was particularly strong – and spilled over to the public domain – in the majority of cases that represented a matter of 'strategic importance'. For example, following a series of demands from Moscow, Buryatiya had to revoke the nomination of Putin's political adversary Yurii Skuratov as its senator in January 2002.[40] A virtually identical situation occurred with the candidate from the Leningrad region, a famous entrepreneur and Yeltsin-era politician Alfred Kokh.[41]

More importantly, the law 'On the status of the member of the Federation Council and status of the deputy of the State Duma' was designed to enable the Kremlin to exert further influence on the composition of this institution. The document granted the Council additional authority to dismiss its senators in cases involving some gross violation of the chamber's political and ethic codes. In order to initiate a dismissal, the Council must obtain just one fifth of the majority vote.[42] Despite the fact that it afforded the Council members more authority, this amendment has opened the way to the politicised removal of 'inconvenient' figures from the institution. The expulsion of Andrei Vikharev (May 2003), who wished to run for gubernatorial elections in Sverdlovsk Oblast without the approval of the Kremlin, and Vasilii Shakhnovskii, a *Yukos* executive under investigation for tax evasion (November, 2003), were cases in point. Moreover, the law on senators broke the links between the regions and the Council by demanding that the former seek FC speaker approval in dismissing the upper chamber deputies.[43] Thus, the change of power in the regions does not involve the change of senators, which moves the FC deputies away from a close participation in regional politics.

The Kremlin's interference into the Council formation process has had its negative effects in the long term. By 2007 it resulted in a situation in which the political links between the senators and their regions had become broken. This process had a gradual character. Indeed, according to statistics published in December 2001, 23 per cent of new senators were former heads of regional legislatures and leaders of important legislative committees, some of whom fled their positions purposely to represent their respective regions in the Federation Council. Eleven per cent comprised high local officials such as vice-governors or deputy heads of regional governments.[44] This tendency indicated the existence of a strong link – even continuity – between the new senators and regional political elites. However, in his final May 2007 address to the Federal Assembly, Putin admitted that up to 70 per cent of the senators came from Moscow and had no association with their territories.[45] More importantly, the June 2007 regulations introducing a 10-year regional residency requirement for the potential senators have not resulted in a general rotation of the current Council membership.[46] Neither have these changes altered the pre-existing centralising appointments vector. The Kremlin still forced Arkhangelsk region Duma to appoint Russia's ex-Minister of Internal Affairs Vladimir Rushailo as its senator in December 2007.[47] Given that the senators cannot fulfil their regional representative functions, the influence of the upper chamber on Russia's legislative process has become minimal. Remington (2003, pp. 680–3), in his quantitative research of the FC voting patterns, notes a dramatic decline in the number of votes that went against the Kremlin's proposed policies.

Migration of cadres

A lack of transparency and reciprocity was particularly visible in the centre-regional migration of personnel. The Kremlin attempted to strengthen control over the most independent regions by 'neutralising' political opponents within these areas before including them in federal power structures. For example, after the resignation of the president of the republic of Ingushetiya Ruslan Aushev on 27 December 2001, who was too outspoken in his criticism of the war in neighbouring Chechnya, many thought that this step was part of a compromise with the Kremlin. Despite Aushev's repeated claims that his resignation aimed to keep Ingushetiya from holding presidential elections at the same time as the parliamentary race scheduled for March 2002, it seemed coincidental that less than two weeks later his interim successor appointed Aushev to represent the republic as a senator in the Federation Council.[48] Similarly, the centre exploited all legal opportunities to prevent the increasingly independent president of Yakutiya Nikolaev from running for a third gubernatorial term. At the same time, however, the Kremlin welcomed his election to the Federation Council and the pro-presidential head of the upper chamber Mironov promoted Nikolaev to the post of vice speaker.[49] The former governor of Primorsk Krai Evgenii Nazdratenko was forced to leave his position in February 2001 but shortly afterwards was appointed to head the State Fishing Committee (Kovalskaya 2001). The June 2003 'promotion' of St. Petersburg ex-governor Vladimir Yakovlev to a ministerial position in the federal government was a result of Putin's desire to replace a former political adversary, Yakovlev, with a more loyal candidate Valentina Matvienko (Hahn 2004, pp. 115–25).

The influx of lower-ranking politicians from St. Petersburg to the highest positions at the federal centre exacerbated the atmosphere of arbitrariness in centre-regional cadre migration. In most cases, such appointments were arranged solely on the principle of personal familiarity with the president and disregarded important factors such as the degree of partisan or political participation. The rapid ascent of Boris Gryzlov provides a good example. Gryzlov, who was formerly employed in one of St. Petersburg's research institutes, was given an unprecedented promotion in March 2001 by being appointed to head the Ministry of the Interior. In November 2002 he was further promoted to lead the *Edinaya Rossiya* party, which made him the speaker of the State Duma following the December 2003 parliamentary elections. The career of Sergei Mironov is also noteworthy. A former official in the St. Petersburg mayor's office under Anatolii Sobchak, Mironov was later given the post of Head of the Federation Council. Likewise, Viktor Cherkesov and Georgii Poltavchenko, who both worked in the St. Petersburg KGB during the Soviet period, became presidential representatives in the North-Western and Central Federal Districts, with Cherkesov moving to Moscow in 2003 to head the anti-drug committee of the Federal Security Service.[50] Equally important remains the fact that Putin actively promoted representatives of legal enforcement agencies (*siloviki*), among which he, as a former director of the FSB, had much weight, to important political positions at the nation level (Colton *et al.* 2005, pp. 10–11 and pp. 24–5). Kryshtanovskaya and White (2003, p. 293) estimate that, while the Yeltsin elite was just 6.7 per cent composed of the *siloviki*, the figure had soared to 26.6 per cent by the end of Putin's first term. The mere influx of *siloviki* to national institutions does not automatically lead to a 'militarisation' of politics, as seen in the poor performance of a large number of such politicians (Gaman-Golutvina 2004). However, such dynamics cannot help but create an atmosphere in which the emphasis is placed on hierarchy and exclusivity at the expense of transparency and reciprocity.

Furthermore, a range of legislation adopted during Putin's first and second presidencies deprived regional politicians of the incentives to migrate to the national level and thereby encroached upon the consolidation of central and regional political elites. Allowing a number of governors to stand for a third term, enacted on 4 July 2003, solidified the composition of the existing regional elites and impeded the emergence of new politicians of both regional and federal importance. Moreover, this policy motivated regional leaders to pursue their careers at the regional level only. The former speaker of the State Duma Gennadii Seleznev observed in spring 2003 that the most independent and capable regional leaders – such as Viktor Ishaev from Khabarovsk Krai, Alexandr Tkachev from Krasnodar Krai, Eduard Rossel from Sverdlovsk Oblast, Egor Stroev from Orel, Mintimer Shaimiev from Tatarstan and others – often chose to remain within their regions rather than proceed to national politics by accepting offers of promotion from the federal centre.[51] This behaviour followed a simple logic. The right to stand for a third, and in some cases fourth term, along with the possibility of the direct appointment by the president, enabled the heads of the regional executives to hold maximum authority within their territories for an indefinite period. At the same time, employment at the federal centre, invariably associated with intrigues and elite reshufflings, carries a number of inherent dangers that are otherwise absent at the regional level. The initiative to cancel direct popular voting for the governors in favour of virtual appointment by the president exacerbated this trend. We shall discuss these developments at length in Chapter 6 of this volume.

Finally, the system of formation of the Federation Council also prohibits effective migration of regional cadres to the national level. The discussion above has demonstrated the role of the federal executive in this process. In addition, the reform enabled prominent business executives who were 'friendly' to the governors to dominate the Council's floor. Statistics published in December 2001 demonstrate that the influx of business elite to the Council was impressive: as much as 34 per cent of this body was composed of representatives of financial and industrial circles.[52] Virtually all Russia's large holdings and major energy companies secured their 'own' senator in the Federation Council. Some of these nominations were pushed through by capitalising on various economic weaknesses of the regions. For example, the legislative assembly of Saratov Oblast approved the nomination of Valentin Zavadnikov, a Muscovite, deputy head of Russia's electricity monopoly *RAO UES*, and closest associate of Anatolii

Chubais, due to the imminent energy crisis in the region. In exchange for this nomination, *RAO UES* cancelled the region's debts to the company for its electricity supplies.[53]

Parties

The initial positive effects at national party building were marred by attempts by the federal centre to institute national parties 'from above' and by the deployment of administrative mechanisms by the regional leaders. The establishment of the Kremlin's personal party fell under the general dimensions of Putin's centralising policies and broadened his popular and institutional power base. Hale *et al.* (2004, p. 295) estimate that only 9 per cent of the *Edinaya Rossiya* electorate cast their vote for the party during the December 2003 campaign on the basis of its previous work in the national legislature, while as many as 26 per cent of the voters supported *Edinaya Rossiya* following Putin's recommendation. A similar situation occurred during the December 2007 campaign, when Putin decided to head the party electoral lists in order to prevent the potential serious loss of image caused by gaining only 40–45 per cent of the vote.[54]

Moreover, that the use of administrative mechanisms, through which the regional governors could influence electoral outcomes, impedes the formation of an effective party system in Russia has long since become an academic contention. Hale (2003) details the means and origins of such electoral 'machines', while Stoner-Weiss (2002, pp. 141–3) insists that economic and bureaucratic interest groups involved in electoral processes are not interested in creating strong parties which could subsequently check their political activity. Golosov (2003, pp. 5–6) supports this view by suggesting that the abundance of financial and administrative mechanisms of a non-partisan nature deprives Russia's regional politicians of the need to create an effective integrated party system. The December 1999 parliamentary campaign particularly demonstrated that party formation has been largely reliant on the administrative machinery and not on grass-roots movements. The competition between the *OVR* (Fatherland-All-Russia) party, led by the leaders of three of the most important regions (the republic of Tatarstan and the cities of Moscow and St. Petersburg), and the Kremlin's new party of power *Edinaya Rossiya*, which consolidated the support of some 39 governors, shifted the voting pattern of Russia's electorate from an 'urban-rural' spilt to a pro-/anti-Kremlin dimension. This was largely explained by an extensive usage of administrative mechanisms by the regional governors and not by a genuine pro-/anti-Kremlin public preference (Hesli and Reisinger 2003, pp. 145–6).

This trend was continued during the December 2003 campaign. The administrative support for Putin's *Edinaya Rossiya* was overwhelming. The national republics, where the administrative potential of the governors is traditionally strong, ensured the highest number of votes for the party. In Chechnya, Mordoviya, Tyva, Kabardino-Balkariya, Tatarstan, Dagestan, and Bashkortostan more than 60 per cent of the vote was cast in favour of *Edinaya Rossiya*. On the other hand, in the regions where the electorate traditionally took greater interest in political developments, the 'party of power' had around 30 per cent (Turovsky 2004). Moreover, a large number of federal ministers and officials were included in the party's electoral lists: Minister of the Interior Boris Gryzlov, Minister of Emergency Situations Sergei Shoigu, Minister of Industry Ilya Klebanov, and the deputy head of the Ministry of the Interior Vladimir Vasilev to name a few.[55] Naturally, none of these officials intended to take up their legislative mandates, which were subsequently transferred to a number of little known party members who would have not been able to compete with an equal degree of success during the election. The December 2007 campaign exacerbated this picture. While only 31 elected ER deputies refused their mandates following the December 2003 campaign, as many as 118 members declared such intentions in December 2007.[56] It does not come as a surprise that all those officials held high-ranking administrative positions in their territories that were so much needed for running the current electoral race successfully. Among those were the governor of the St Petersburg Federal City Valentina Matvienko, the governor and mayor of Nizhnii Novgorod Valerii Shantsev and Vadim Bulavinov respectively, the governor of the Kemerovo region Aman Tuleev, the governor of the Chelyabinsk region and the mayor of Chelyabinsk Petr Sumin and Mikhail Yurevich respectively, the governor of the Novosibirsk region Viktor Tolokonskii and many other high-ranking regional officials. Moreover, ER has officially promoted those governors who ensured the party's highest voting result in their territories. These leaders were included in the party's national Supreme Council. Just like during the December 2003 campaign, this primarily concerned the most authoritarian regions, where the usage of non-democratic politics and administrative resources was particularly widespread. Among those were the president of Chechnya Ramzan Kadyrov, where ER attracted as much as 99.36 per cent of popular support, the president of Ingushetiya Murat Zyazikov, who ensured 98.72 per cent of the vote, the president of Kabardino-Balkariya Arsen Kanokov (96.12 per cent), the president of North Ossetiya Taimuraz Mamsurov (71.6 per cent), the president of Adygeya Aslan Tkhakushinov (70.79 per cent), the governor of Tyva Sholban Kara-ool (89.21 per cent), and the governor of the Penza region Vasilii Bochkarev (70.31 per cent).[57]

It is also important that the newly adopted partisan legislation had a range of provisions which facilitated the deployment of administrative resources. The law 'on political parties', for example, encouraged these structures to seek private connections with big business and the federal government by demanding the registration of 50,000 members and the establishing of branches in more than one half of the Federation's subjects. It is in this context that Irina Khakamada, a former leader of the Union of Right Forces, declared that the 'party-selling' business has become normal practice in Russia and that she was personally offered a ready-made organisation for the price of just US$1 million.[58] It is significant that the US$1 million price stayed stable for some years, which was confirmed by tycoon Alexandr Lebedev and the USSR ex-president Mikhail Gorbachev in autumn 2008, when they embarked upon building an independent democratic party project.[59] Therefore, it is not surprising that the party registration process saw the emergence of numerous little-known movements with no coherent organisational or ideological links with federal politics. The vast majority of the 48 parties which had been registered by the Ministry of Justice by March 2005 comprised such formations.[60] Moreover, the law on proportional representation, passed by the Duma in its first reading on 24 December 2004, stipulates the abolition of single-mandate districts for elections to the national parliament and introduces a 7 per cent barrier to the parties entering the State Duma.[61] Despite the fact that this document aims at the reduction of the influence of local economic elites on the elections, it is likely to increase arbitrariness in the process of party formation. Given that Russia's contemporary parties largely represent the Kremlin's creations, this law gave the green light to these organisations and precluded possibilities for building genuine parties 'from below'.

Furthermore, the initial attempts to enhance the political role of the parliament have been refuted by the administrative monocentric developments. The overwhelming victory of the *Edinaya Rossiya* party during the 2003 parliamentary election effected a serious blow to the idea of a strong and independent legislature. The absence of a credible opposition in the 4th Duma carried an inherent danger of over-centralisation and the domination of parliament by the federal administrative elite. Indeed, decision-making became the sole prerogative of the Kremlin, thus discrediting legislatures' political image and impeding the development of national parties. Between 2003 and 2007, the lower house of Russia's parliament approved almost all initiatives proposed by the presidential administration without showing any signs of independent functioning. This concerned both economic and political matters. In the economic sphere, the fourth Duma enacted a wide range of measures that institutionalised the system of state capitalism in Russia. These included the adoption of the three-year budget in 2007, the launch of the State Investment Fund in 2005, the creation of the Russian Venture Capital Foundation (April 2006) and the Bank of Development and International Investments (June 2007), and the establishment of a number of state corporations in the aviation and ship-building sectors, as well as in the nano (July 2007) and atomic technologies (March 2004) industries. In the political sphere, the 4th Duma adopted an unprecedented number of restrictive laws that complied with the wishes of the presidential administration. The law on extremist activity (2007) has been amended to include various types of independent political engagement. The new legislation restricting NGO political activity was passed in May 2006, disregarding the wide-scale international and public outcry. The single mandate electoral system had been abolished for the elections to the national legislature, and the right to vote 'against all' candidates has been taken away (2006). The new version of the law on referendum, which significantly restricted the spectrum of questions that can be addressed during such plebiscites, was passed in the first reading in spring 2007.[62]

Such overt compliance with the wishes of the presidential administration tarnished the public image of the State Duma and precluded it from taking part in the party-formation project. Indeed, the December 2007 VTsIOM data sends some alarming signals to the occupants of the lower house of the Russian legislature. The State Duma appeared to be the least trusted and the least approved institution in Russia. Only 28 per cent of the respondents approved its activity against 82, 47, and 41 per cent of those who approved the functioning of the president, Prime Minister, and the government respectively. Similarly, 48 per cent of the respondents thought that the State Duma does not execute its duties appropriately against 37, 33, 26, and 11 per cent of those who thought the same of the government, the Federation Council, Prime Minister, and the president respectively.[63]

Finally, as a result of central interference in the party-building process, regional elites lost any particular ideological attachment to the federal centre and have become guided by pragmatic profiteering considerations in their partisan preferences. Despite the central efforts at promoting a national ideology at the regional level, the regions' commitment to one or another idea remains exceptionally low. Instead, control over economic and political resources determines the course of almost all political struggles evolving in the regions. Most representatives of regional elites have changed party alliegiance a significant number of times in line with various political shifts at the national and regional levels. Clearly, ideology played no role in such decisions. For example, following Putin's announcement to lead United Russia during the December 2007 parliamentary campaign, many members of Just Russia in Voronezh, Sakhalin, Kirov, St Petersburg and other regions left this organisation in favour of United Russia.[64] This took place despite a range of profound ideological differences between the two organisations. Interestingly, following the Just Russia success during the December 2007 parliamentary election, many regional party branches began to consolidate again. In particular, in the Kirov region, the party leaders publicly apologised to the national party leadership for their previous defection.[65] The reconstitution of gubernatorial elections has played its part in such developments by ultimately reducing electoral activity across the land and accelerating the reliance on the administrative resources of the centre by the regional governors (Colton *et al.* 2005, p. 24; Hale *et al.* 2004, p. 315). This favours *Edinaya Rossiya* at the expense of other national parties: over 20 regional governors had joined *Edinaya Rossiya* by March 2005 alone, i.e. within the first three months of the system functioning (Lysenko 2005). Many such actions took place as a result of gubernatorial decisions to renounce their previous party affiliations virtually 'in exchange' for an extension of their terms in office. This trend was particularly visible amongst the Communist and former *Rodina* governors – Alexandr Mikhailov of Kursk Region and Georgii Shpak of Ryazan Oblast are cases in point.[66]

Potential implications and consequences

Our discussion has so far indicated that harmonious federal relations require two main factors: institutions and commitments. It has also become clear that supporting institutions of federal integration in Russia are gradually emerging. However, the commitments are still lacking. The Kremlin, while attempting to establish the essential federal networks, was unwilling to make sure that they function in a manner consistent with the federal values of reciprocity and inter-dependence. Thus, it becomes clear that the main problem of Russian federalism stems from a serious deficiency in processes, and therefore, is cultural and ideo-logical. The style with which both the Kremlin and the regional leaders operate a range of Western-borrowed structures has proven to be informal, arbitrary, and conformist. This situation generates a range of important questions. What impact might the Kremlin's centralising stance have on the regions' perceptions of the federal idea and the federation in Russia? Moreover, if the regions are not pro-fessing federal values, just like the centre does not, what are the potential con-sequences of this double commitment problem for the future of Russian federalism?

Experiences of failed, incipient, and successful federations point to the para-mount importance of cultivating the ideas of reciprocity and interdependence in the working of federal networks. Those countries which overstated the role of the federal executive in the functioning of federal networks encountered political difficulties, while those which cultivated reciprocity and interdependence exhibited substantial progress on the road towards federalism. For example, the maladies of the Nigerian federation are seen in that the essential federal architec-ture, aimed explicitly at accommodating various regional cleavages, national political parties, and migration of personnel, has been vitiated by the overwhelm-ing role of the centre and its potential to redistribute national income (Suberu 2001, pp. 111–40 and pp. 47–79). At the same time, Belgium and Spain actively cultivate a commitment to federal values through the functioning of intergovern-mental networks. The pressures of European integration, in particular in the wake of the Maastricht Treaty, not only contributed to the emergence of a multiplicity of such bodies (Börzel 2000) but also enabled the regions to speak 'as equals' within these venues. This has forced the central and regional governments to co-operate willingly with each other and, arguably, had a restraining effect on separatist tendencies in both countries (Ortino *et al.* 2005, pp. 213–15).

Equally interesting is the experience of India and Indonesia where the cultiva-tion of fairness and mutuality has been made a priority. India enshrined the grant-in-aid system in Articles 275 and 282 of its constitution, which clearly stipulates which states receive these benefits and why. A number of Finance Committees have been established to ensure transparent workings of the system (Agarwal 1959, pp. 12–19; Hicks 1978, pp. 184–5). The Indonesian decentrali-sation process, during which the emergence of centrifugal tendencies in the provinces was a risky possibility, has placed emphasis on institutional transpar-ency in allocating grants. Suggestions have been made to move towards estab-lishing a special Grants Commission that could facilitate this process, making the transfer system more equitable and clear to everyone in the provinces and in the centre (Alm *et al.* 2004, pp. 159–99). This contrasts with the largely arbitrary

Russian approach, in which the centre deployed the federal programmes system as a means of achieving its immediate political objectives within certain strategic regions.

The workings of older federal states are perhaps less relevant to the incipient Russian federalism. However, it can provide some important suggestions of a theoretical value. Smiley (1972, p. 441), while generally praising the development of intergovernmental co-operation in Canada, underscores that this phenomenon was born out of the 'unavoidable interdependence of the responsibilities of the two constitutionally ordained orders of government... which requires a continuous process of federal-provincial consultation and negotiation'. Thus, the emphasis is not on the dominance of a federal level over the regions but on understanding that the complexities of modern societies do not provide for either strict hierarchy or isolation, and therefore require fundamental commitments to reciprocity and interdependence. It is on this basis that Smiley terms the Federal-Provincial Conference of Prime Ministers and Premiers in Canada as 'one of the most crucial elements of Canadian federalism' which, 'in terms of potentiality, could prevail even over the constitution' (Smiley 1972, p. 60). Indeed, while the Canadian Conference has often had meetings with other important national institutions – the Continuing Committee on Fiscal and Economic Matters and the Tax Structure Committee being the most prominent – Russia's State Council did so with the national Security Council, indicating the predominant role of the federal executive. The regions' voice in the Special Premiers' Conference of Australia has also been strong. In November 1991, the States even announced a collective boycott of the meeting with the Commonwealth Prime Minister and convened an independent conference in Adelaide (Galligan 1995, p. 208) – a development unthinkable in the Russian case.

The emphasis on reciprocity in centre-regional cadre migration also contributed to a varying degree of success across federations. Ordeshook (1996, pp. 207–10) claims that centrifugal tensions which bedevil the Canadian federation, in contrast to the relative stability of the United States and Germany, stem largely from the fact that the most gifted, capable politicians tend to remain at the local level. Belgium attempts to cultivate federal commitments by instituting a mixed composition of regional political offices – regionalists who wish to pursue their political career at the sub-national level only are balanced by federalists, 'who wish to move up to the national level, including the prime ministership' (Covell 1987, p. 73).

Furthermore, from a comparative point of view, the weakening on Russia's bicameralism brings into question the stability of Russia's federal system. Lijphart (1999, pp. 202–3), in his study of 36 democracies, contends that the world's most stable federations, such as Australia, United States, Germany and Switzerland, tended to have strong bicameral structures. Medium-strength and weak bicameralism, on the other hand, has been attributed to some less stable federations and unitary states experiencing a complex process of regional devolution. Canada, Belgium, Venezuela, Spain, Italy, and United Kingdom have fallen into this league.[67]

This brief comparative outlook indicates that ideological and cultural limitations of Russia's federal system do not signal an optimistic message for the country's political dynamics. The overwhelming predominance of the central state in the functioning of centre-regional institutions precipitates regional leaders' pliability to the Kremlin's wishes. At the same time, it is clear that the origins of such compliance are highly dubious. They stem, first, from the regions' desire to secure immediately realisable practical advantages and, second, from their fear of administrative sanctions.

With regards to the first part, many regions followed this behavioural pattern. In 2005, Russia's Finance Minister Alexei Kudrin openly admitted that the centre could not have conducted the regional plans unilaterally and 'purchased' the loyalty of many key regions during the early stages of the federal reform (Turovsky 2005). For example, the governor of Khabarovsk Krai Viktor Ishaev supported all presidential policies on the reaffirmation of central state power, and then demanded an increase in governmental orders for the region's military enterprises and the expansion of subsidies for the development of transportation. Similarly, the ex-governor of Krasnodar Krai Nikolai Kondratenko forwarded a number of memos to Putin in support of the president's initiatives on reforming the region's agricultural sector. He then requested that the centre channel more funds to the regional budget.[68] Arguably, placing such demands is not an unusual practice. Counterparts of Russia's regional leaders in other federations are also engaged in distributing struggles over fiscal allocation, jobs, tariffs, cabinet appointments, civil recruitment and the like. These claims represent, according to Rothschild (quoted by Suberu 2001, p. 9), 'negotiable' demands of territorial units. What seems to be more worrying, however, is that against the backdrop of centralising and irresponsive federal policies, such demands may give way to 'non-negotiable' claims of an ethnic, religious, territorial or cultural character. This, in turn, could be threatening to the very existence of a federation, as the experience of Slovenia, Bangladesh, and some African regions shows (Ortino *et al.* 2005, p. 123; Hicks 1978, p. 173; Franck 1968, pp. 167–83; Suberu 2001, p. 9).

The second part arguably represents a more serious danger signal in that it leads to simmering regional discontent. Indeed, various elite and population surveys on the issues of federalism and regional relations continue to deliver alarming results. In 2001, 14.6 per cent of regional leaders named centre-regional tensions as one important threat to the Russian state (ROMIR 2001). Polls conducted in autumn 2001 with 250 senior officials of regional administrations and governments by the Sociological Research Centre of the Russian State Service Academy under the president of the Russian Federation are even more illuminating. Given that this survey was conducted by representatives of the Presidential Academy – an institution that could deter the respondents from responding truthfully – the answers falling into the 'I don't know' group are the most indicative. These polls have demonstrated a predominantly negative or 'I don't know' attitude towards Putin's leadership. Putin's ability to act democratically fell into the 'I don't know' column in 40 per cent of the cases and was evaluated negatively by 22 per cent of the respondents. The extent of reciprocity and mutuality in his decision-making procedures was placed in the 'I don't know' category by 32.3 per cent of the respondents and evaluated negatively by as many as 27.9 per cent.[69] An extensive range of interviews conducted between 2002 and 2005 with over 174 regional politicians in 6 regions led Russia's most prominent regional sociologists Chirikova (2005, pp. 195–210) and Lapina (2005, p. 74) to point at the existence of a hidden regional opposition. Similarly, polls conducted by the Independent Fund of Public Opinion (FOM 2005) cited the lack of the regional leaders' commitments to the federation as among the main threats to the integrity of the Russian state.

Simmering political discontent, coupled with the fear of administrative sanctions, was well illustrated during the launch of the new gubernatorial elections formula. The former leader of the Yaroslavl region Anatolii Lisitsyn insisted that the regions backed the new system of gubernatorial appointments largely because the centre threatened to withdraw subsidies from those who refused to comply.[70] It was one main reason why the regional assembly attempted but failed to mount opposition to this initiative.

Ever more worrying signals emanate from the public opinion surveys. The December 2003 ROMIR (ROMIR 2003) monitoring data indicates the instability of state power (16 per cent), ethnic conflict (10 per cent) and the situation in Chechnya (11 per cent) as Russia's most serious societal problems. In 2005, the majority of Russians considered that the disintegration of the federation was a possibility. 20 per cent suggested that such an outcome was probable, 32 per cent responded that it was possible, and only 10.8 per cent suggested that this could never happen.[71] Moreover, over a third of Russians (38 per cent) thought that events similar to those in Andijan were possible in Russia with the change of political leadership in 2008.[72]

Finally, and following from above, if the centre holds troublesome regions by redistributing lucrative benefits and the rest of the federation by its overwhelming political might, then the future of the federation is quite uncertain. It is uncertain in the sense that it depends not on the amalgam of institutions and commitments but on a composition of political actors at the centre (that are currently capable of enforcing their will) and on a particular economic situation (that enables the Kremlin to deliver various subsidies and concessions).

The economic issue seems especially salient in that the extreme dependence of the country's economy on global commodity prices (Colton *et al.* 2005; Hale *et al.* 2004) could, at some point, jeopardise the centre's ability to deliver economic subsidies. This threat is particularly worrying even in a comparative light. The example of the Nigerian federation, which succumbed to civil war following the discovery of oil in the early 1970s and continues to experience oil-revenue redistribution problems (Hicks 1978, p. 173; Suberu 2001, pp. 111–40), demonstrates that natural riches might at some point turn from a blessing to a curse. Another important point has been made by Whitefield (2005). He (2005, p. 144) shows that those who identify with the post-Soviet Russian state belong to the most economically successful and politically powerful social class. He, therefore, argues that the growth in Russian identification is largely dependent not on shared conceptions of political freedom but on economic performance that may decline. Similarly, any political instability related to a change in the composition of central elites may well lead to a resurgence of regional claims for further autonomy, sovereignty or financial independence, the memory of which is still fresh from the Yeltsin era.

5 Business and politics in Russia's regions

The problem of political style

In this chapter we will examine the impact of Russia's *political style* on the development of business-state relations within the regions and the subsequent influence of these dynamics on the evolution of centre-regional political dialogue. This discussion will examine the development of regional institutions and the degree of their autonomy (or otherwise) from the Kremlin and differing influence groups of national importance. The chapter will argue that Putin's institutional reforms at the centre conducted during the initial stages of his presidency have had a significant effect on the redistribution of power within regional structures. The analysis will concentrate on the financial-political aspects of regional institutional development. It will be demonstrated that the centre's attempts to change the rules of national political interaction by introducing a number of steps to dismantle the system of oligarchic monopolism at the federal centre immediately affected the regional establishment. The political system, under which representatives of large corporations had a serious influence on policy-making, migrated from the national arena into the provinces and significantly altered the influence of previously powerful regional leaders. By the end of Putin's first term, these dynamics had resulted in the establishment of effective non-institutional relationships between big business and Russia's regional authorities. The emergence of poorly controlled oligarchic structures at the subnational level prompted the centre to adopt a range of initiatives aimed at placing a safe and controllable distance between representative of nationwide enterprises and regional governors. With the exception of changes that took place in the regional electoral systems, such policies have been predominantly borrowed from well-functioning Western models. At the same time, we will argue that the *political style* and context within which such ideas have been implemented have subverted the initial intentions and led to the emergence of alternative, and at times more sophisticated, forms of relational informality.

The emergent problems and proposed solutions

During the first term of Putin's presidency, business and state in Russia's regions have been forming an increasingly symbiotic, interdependent, and mutually beneficial relationship. Academic experts (Turovsky 2001; Turovsky 2002;

Zubarevich 2002; Orttung and Reddaway 2004; Frukhtman 2005) agree that, apart from the objective economic factors, such as a full-scale recovery of Russia's industrial and resource extracting capacities, this dynamic has largely represented a 'side effect' of two key developments in central policy: (1) de-oligarchisation of the federal centre, and (2) efforts at re-centralisation of regional politics.

First, with Yeltsin's departure from office, the Kremlin gradually established a dominant position in its relations with economic elites, whose influence on central political institutions markedly diminished. Putin's anti-oligarch campaign, which had its peaks during May-June 2000, during the first half of 2002, and during autumn 2003 led to the exclusion of a large number of business executives from the Kremlin and placed them at a certain 'safe' distance from the epicentre of decision-making processes.[1] This situation left Russia's economic elites with little choice but to explore political opportunities in the regions and to embark on a struggle for influence within provincial institutions.

Second, the penetration of the regions' administrative structures by big business was facilitated by the fact that, prior to the *Yukos* affair, which was launched during the second half of 2003, Russia's tycoons were not part of the executive 'power vertical' pursued by president Putin. It has therefore become a common agreement that, due to the fact that Russia's leading industrialists managed to retain their position within the existing power elite for most of Putin's first term, they had a comparative advantage over the governors in the struggle for influence in regional politics (see Orlov 2002; Zudin 2002; Turovsky 2002). Therefore, the central political campaign, which aimed to trim back the authority of the governors and incorporate sub-national leaders into the presidential 'power vertical', weakened the political potential of regional administrations and facilitated the rapid and competitive expansion of big business into regional institutional structures. In other words, the exclusion of representatives of large corporations from the Kremlin, combined with the launch of centralising reform, left sub-national politics increasingly vulnerable to the influence of industrial conglomerates of national importance.

Indeed, following the launch of Putin's reforms, Russia's integrated business groups (IBGs) began to target sub-national administrative apparatuses. Because such financial groupings represented the main contributors to regional budgets, their boards of directors sought to place representatives in leading positions in executive and legislative bodies in the federation's subjects. This was done with the aim of manipulating regional financial and legal resources to their advantage. Until the end of 2003, when relations between big business and power began to change drastically, the financial group created by President Yeltsin – commonly known as 'the family' – represented one of the most successful business lobbyists both in the regions and in the centre. With the departure of Berezovskii from key positions in the clan in 2000, Roman Abramovich, the former owner of *Sibneft*, assumed full financial and political power within the coalition; he was later joined by Oleg Deripaska, the head of one of the world's largest aluminium companies *Russian Aluminium*.[2] 'The family's' consolidated business positions in the regions included *Sibneft, Russian Aluminium, Ruspromavto*, the *Ural Metallurgy Company, Eurasia-Holding*, the *MDM* group, *Aeroflot, Rosno Insurance, GAZ*, and *Pecherugol*.[3] The aluminum industry represented the most successful sphere of the company's activity, in which it managed to secure a virtual monopoly. Having no scope for further industrial expansion by the middle of the 1990s, aluminium executives embarked on a struggle for administrative-political influence in the regions which hosted the company's plants. For example, 'the family'-sponsored General Lebed won the gubernatorial election in Krasnoyarsk Krai in 1996 and placed an ex-manager of *Russian Aluminium (Rusal)* Nikolai Ashlapov in the post of first vice-governor of the region. During the period 2000–2, 'the family' was similarly successful in obtaining political control over Khakasiya's government (Makarkin 2002). Despite their defeat in the gubernatorial elections in Krasnoyarsk Krai in October 2002, the post of head of the regional assembly was occupied by 'the family'-sponsored candidate Alexandr Uss.[4] In the oil industry, the ex-owner of *Sibneft* Abramovich won the gubernatorial elections in Chukotka on 24 December 2000. Moreover, the governor of Omsk region Leonid Polezhaev had a close financial relationship with Abramovich and was often considered a promoter of *Sibneft*'s interests (Kravtsov 2003). The financial segment of 'the family' – the *MDM*-group – managed to consolidate its political positions in major Russian coal-mining regions. Komi Republic was one of the most illuminating examples where the regional administration overtly represented the interests of this corporation.[5] In addition, the head of the board of directors of one of MDM's regional branches Alexei Lysyakov became a senator for Stavropol Krai, where the group has strong interests in the chemical industry.[6]

Other financial holdings such as *Alfa-TNK, Lukoil, Yukos, Gazprom-Sibur, RAO UES, Surgutneftegaz, Norilsk Nikel-Interros, Mezhprombank*, and *Severstal* could not compete with 'the family' in terms of regional expansion. Nevertheless, the influence of *Lukoil* in Volgograd Oblast extended to the level of vice-governor Yurii Sizov, a direct company representative, and in the Nizhnii Novgorod region to minister of finance Vadim Sobolev.[7] *Yukos* gained control over Evenkiya in April 2001 by electing its ex-executive Boris Zolotarev to the post of regional governor. Similarly, Tomsk Oblast was always regarded as a *Yukos* fiefdom and the governor, Viktor Kress, a company loyalist (Kravtsov 2003). The corporation also secured influence in Mordoviya by appointing its foremost executive Leonid Nevzlin to represent the region's administration in the Federation Council in November 2001 and substituting this nomination in March 2003 with the head of *Yukos-Mordoviya* Nikolai Bychkov. *Norilsk Nikel-Interros* secured solid positions in Krasnoyarsk Krai and Taimyr Autonomous Okrug by electing Alexandr Khloponin (on 3 October 2002) and Oleg Budargin (on 26 January 2003) respectively to the posts of regional governors.

Very often regional gubernatorial elections became a venue for corporate contests, in which one group or another supported their candidate for the highest post in the region. During the gubernatorial elections in Nenetsk Autonomous Okrug in January 2001, *Lukoil* candidate Alexandr Shmakov ran against

Severnaya Neft representative Vladimir Butov.[8] In Tyumen Oblast in January 2001, Sergei Sobyanin supported by *Sibneft* and *Transneft* faced Leonid Roketskii of *TNK*.[9] The gubernatorial elections in Irkutsk Oblast, which took place in July–August 2001, witnessed a struggle between Boris Govorin supported by *Rusal* and *Alfa* and Valentin Mezhevich, who was nominated by *Interros*. The interest shown by these companies in the region was mainly determined by the launch of a contest to manage the Kovyktinskii gas well.[10] Moreover, immediately following his victory governor Govorin placed *Rusal* executive Vladimir Kolmogorov at the head of the region's largest electricity company *Irkutskenergo*, thus depriving the centre and *RAO UES* of control over one of its best electricity assets.[11] In Magadan Oblast, differing business structures supported their own candidates during the March 2003 gubernatorial elections: the Mayor of Magadan Nikolai Karpenko (who was also promoted by the federal centre and subsequently lost the race in the second round) and deputy governor Nikolai Dudov.[12] In December 2003, former *Nornikel* executive Dmitrii Zelenin defeated Igor Zubov, a representative of another nation-wide business structure *AFK-Systema*, in the gubernatorial elections in Tver Oblast.[13]

At the end of 2002, the former Minister of the Antimonopoly Policy Committee Ilya Yuzhanov publicly acknowledged the fact that financial business groups had managed to secure tight control over the socio-political affairs of most regions. More importantly, the Minister added that, in order to avoid competition and conflicts, IBGs had informally agreed to operate in separate territories, thus transforming certain regions into their personal fiefdoms and creating virtual monopolies within given areas. For example, oil industrialists divided their spheres of influence in accordance with a territorial principle: 13 zones had been created with the predominance of one corporation. A number of regions that were not engaged in oil production were taken over by large companies which represented other sectors of the economy such as metallurgy, coal-mining, forestry, production of chemicals, etc.[14] In 2003, the proportion of business executives in regional governorships reached 14.7 per cent (Kryshtanovskaya 2003, p. 32), and by 2005, as many as 16 of Russia's regions (approximately 18 per cent) were headed by top executives and owners of large corporations (Lapina 2005, p. 74).

The following table and map illustrate the penetration of Integrated Business Groups into certain resource-rich regions of Siberia and the Russian North and their official power resources in regional administrations at the beginning of 2003.

Sociologically, this process was reflected in opinion polls: by the end of 2002, there was a remarkable decline in the regional governors' power rating compared to the Yeltsin years and a substantial increase in a similar rating of big business executives, who had infiltrated and influenced sub-national administrations. Monthly reports of the Fund of Public Opinion (FOM), published in *Nezavisimaya Gazeta*, continually cited prominent industrial leaders in lists of the most influential lobbyists and politicians of the month.[16] Until the launch of an open anti-oligarchic campaign in October 2003, the lists included names such as

Table 5.1 Penetration of IBGs into West Siberian and Northern Economic Districts[15]

Region	Financial group	Group's representative in regional administration
Chukotka Autonomous Okrug	Sibneft & MDM	Governor Roman Abramovich, owner of *Sibneft*
Evenk Autonomous Okrug	Yukos	Governor Boris Zolotarev, former *Yukos* executive
Irkutsk Oblast	Rusal	Governor Boris Govorin; the gubernatorial campaign was financed by *Rusal*
Kemerovo Oblast	MDM	Governor Aman Tuleev advocated the group's interests in the region
Khanty-Mansi Autonomous Okrug	Lukoil & Surgutneftegaz	Governor Alexandr Filipenko; the gubernatorial campaign was supported collectively by the two companies
Komi Republic	MDM	Nikolai Levitskii, deputy governor, ex-manager of Evrokhim. Governor Vladimir Torlopov
Krasnoyarsk Krai	Interros- Nornikel	Governor Alexandr Khloponin, *Nornikel* executive
Nenetsk Autonomous Okrug	Severnaya Neft	Governor Vladimir Butov, owner of *NNK*; the gubernatorial campaign was financed by *Severnaya Neft*
Omsk Oblast	Sibneft	Governor Leonid Polezhaev, deputy governor Andrei Golushko
Republic Khakasiya	MDM	Governor Alexei Lebed; the electoral campaign was financed by the *MDM* group
Republic Tyva	Mezhprombank	Senator Sergei Pugachev, the head of Mezhprombank
Taimyr Autonomous Okrug	Interros- Nornikel	Governor Oleg Budargin, *Nornikel* executive
Tyumen Oblast	*Until January 2001* TNK; *then* Sibneft	Governor Leonid Roketskii, former head of the board of directors of *TNK* lost to *Sibneft* representative Sergei Sobyanin
Tomsk Oblast	Yukos	Governor Viktor Kress; the gubernatorial race was financed by *Yukos*

Roman Abramovich (*Sibneft*), Vagit Alekperov (*Lukoil*), Petr Aven (*Alfa Group*), Oleg Deripaska (*Bazel*), Mikhail Fridman (*Alfa Group*), Mikhail Khodorkovskii (*Yukos*), Alexandr Mamut (*MDM Group*), and Vladimir Potanin (*Interros*). Interestingly, these individuals were always placed above or among key figures of Putin's presidency – Sergei Mironov, Dmitrii Kozak, Vladimir Ustinov, Sergei Ivanov – and at all times above both the heavyweight governors

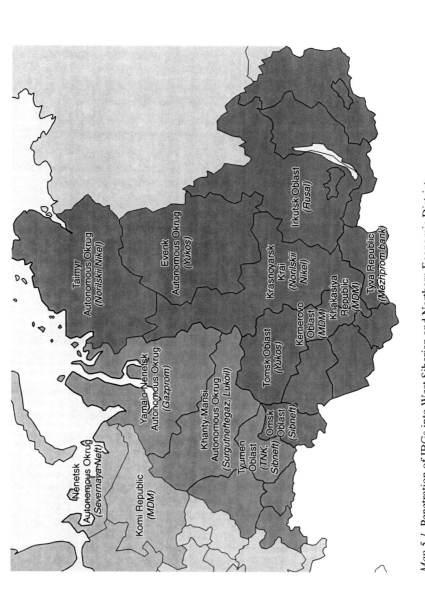

Map 5.1 Penetration of IBGs into West Siberian and Northern Economic Districts.

of Yeltsin's era and important regional figures in Putin's Russia such as Mintimer Shaimiev, Egor Stroev, Eduard Rossel, Aman Tuleev, Boris Gromov, Alexandr Khloponin and all seven plenipotentiary representatives.[17] Even in October 2003, when the analysts drafted their study prior to the arrest of the head of *Yukos*, Khodorkovskii and Abramovich were ranked 5th and 6th respectively. This was above Ministers Alexei Kudrin, Boris Gryzlov and Sergei Ivanov, and above the key figure of the presidential administration Vladislav Surkov.[18] This situation only changed in November 2003, which saw the rise of *siloviki* to the highest positions in the political hierarchy.[19] It therefore became apparent that the oligarchic monopolism established at the federal centre during the second half of the 1990s had managed to re-consolidate its forces and institutionalise its positions in the regions by the second half of Putin's first term. It was in this connection that Zudin (2002) insists that such a system became 'the most successful model of institutional development in Russia's regions and represented the primary arrangement influencing the dynamic of centre-regional relations'.

The entrenchment of the oligarchic monopolism system in the regions has led to a series of negative social developments, which I will continue to discuss throughout this chapter. Here, I shall briefly outline some of the major traits of these dynamics. First, the system has diminished the potential for the advancement of high technologies and refining industries into the regions. This primarily concerns the resource-rich territories, which have represented common targets for the oligarchic system of governance because they can generate quick income from minimal non-technological investment. Second, these developments have led the regions to a situation in which the distinction between business and power has become blurred. This largely restricted the independence of regional executives and limited their ability to address the most pressing needs of their electorates. Consequently, the transparency and legitimacy of sub-national administrative systems, as well as the accountability of their leaders, have developed into serious socio-political problems. The governors have often pursued the interests of those corporations that brought them to power often at the expense of vital economic interests of their regions. These dynamics resulted in the creation of an artificially high level of entry to regional business for other companies, even very strong ones, which impeded the development of economic competition, the business environment and the establishment of a strong class of middle executives. In a number of cases, the privatisation of regional industrial assets led to ineffective management of privatised enterprises. This situation resulted in a serious decline in the social spheres of the territorial units in which these plants were located. This was primarily associated with the fact that, due to the specifics of the former Soviet economy, for many regions such companies represented principal employers and contributors to local budgets.

The situation in the Komi republic perfectly illustrated this point: economic expansion of the *MDM* group to republican coal mining industries (which comprises 50 per cent of regional budget inflows) enabled *MDM* investors to infiltrate the highest circle of regional authorities.[20] Due to such 'close co-operation', regional leaders permitted *MDM* to privatise regional shares in important mining wells such as *Vorkutaugol* and *Intaugol*, and to avoid fair market competition with other corporations such as *Severstal*, which also displayed interest in the deal.[21] The monopoly of *MDM* over coal mining enterprises led to a decline in production levels at a number of shafts in Vorkuta and Inta. The company ignored the problems of development and modernisation of these shafts and concentrated on the provision of cheaper coal from alternative sources. By the time the region decided to transfer the management rights to potentially more effective *Severstal* in July 2003, the regional budget suffered losses of no less than 350 million roubles. More importantly, as a precondition *Severstal* pressed for the closure of two large mines in Inta and Vorkuta, thus creating the need to relocate some 100,000 workers from the area.[22]

Furthermore, large financial holdings engaged in controversial relationships determined by the logic of their regional expansion. These took the form of co-operation, co-existence, and economic wars. The latter form has become particularly prominent, as it has been associated with increased competition for control over the country's industrial assets and natural resources. Responses by regional governors to the arrival of big business in their territories facilitated the development of such a trend. The resulting economic conflicts threatened socio-political stability at the regional and national levels. In many cases, political assassinations and false criminal prosecutions have become a means of resolving economic conflicts. In this context, the manipulation of local legal systems has developed into a serious social problem. Gubernatorial elections in Krasnoyarsk Krai illustrated this point: in the wake of its apparent defeat in gubernatorial elections, the disaffected 'family' group exerted pressure on the region's electoral committee and persuaded it to annul the election result. Regardless of the fact that newly elected governor Khloponin managed to keep his position with visible support from the centre, the incident markedly diminished his legitimacy. In addition, the virtual monopoly on coal supplies enabled the family-owned *Vorkutaugol* to block deliveries to Krasnoyarsk Krai and by so doing to create economic obstacles for the newly elected governor.[23] Similarly, the arrest on false allegations in July 2001 of Viktor Ledovskii, the former director of Russia's fourth largest producer of chemical fertilisers *Azot*, took place as a result of *MDM*'s semi-criminal struggle to purchase the factory.[24] The murder of the Mayor of Taganrog Sergei Shilo in Rostov Oblast in October 2002 took place against the backdrop of severe conflicts between governor Vladimir Chub and representatives of *MDM* group and the struggle between *MDM* group and *Mezhprombank* for economic-political influence in the region.[25]

In addition, large industrial groups had the tendency to manipulate public opinion via local media. For example, *Russian Aluminium* established a new TV channel in the republic of Bashkortostan, Russia's last non-privatised region. With the aim of infiltrating the region, the oligarchs initiated a media campaign against the regional administration by siding with the centre in a dispute over the constitutional status of Rakhimov's presidency.[26] A similar situation occurred in Irkutsk Oblast. Having secured financial control over Ust-Ilimsk forestry plant in autumn 2001, Oleg Deripaska, supported by a number of regional Duma deputies, launched an open media war against his competitors from the St. Petersburg factory *Ilim Pulp.* [27] The governor of Omsk Oblast Leonid Polezhaev often put pressure on media channels which exposed his close financial relations with *Sibneft.*[28]

Finally, and more importantly, the participation of large corporations in regional politics gradually provided the provinces with nationally important financial patrons and thereby with stronger negotiating positions with the centre. As Orttung and Reddaway (2004, p. 292) observe, 'the combination of regional political power with the economic resources of some of Russia's largest companies could put severe constraints on certain sorts of action by the federal executive'. Therefore, the emergence of poorly controlled poles of political and economic influence in the regions led the federal authorities to alter the mode of interaction with regional leaders at the national level. By the end of his first term, Putin had decided to put the brakes on the rapidly developing business-political collaboration in the regions and make the relationship between the two actors more formal and institutional. Four main policy initiatives have been launched with the view of creating a 'safe' distance between business and politics at the regional level.

First, the thesis on corporate social responsibility (CSR) was introduced. This new political doctrine called all large and medium-sized corporations to take a diligent stance towards their fiscal duties and to go beyond legally stipulated requirements by making some voluntary contributions towards social developments within the regions of industrial coverage.[29]

Second, the system of taxation and licensing has been reformed so as to deprive governors and large corporations of some important mechanisms of political manipulation. In particular, internal tax havens have been outlawed from 2004 with the aim of dampening large corporations' ability to deploy fiscal optimisation mechanisms (Rubchenko and Shokhina 2003). The state has also restricted governors' access to the redistribution of regional mining licenses. This was formalised in the amendments to the law 'On Natural Resources' which came into force on 1 January 2005.[30]

Third, the creation of a wide range of institutions has been actively promoted so as to introduce greater transparency in the relationship between the two parties. Finally, the system of gubernatorial elections was reconstituted in order to minimise the extent of business-political electoral collaboration. The laws 'on organisational principles of executive and legislative bodies of the Russian Federation's subjects' and 'on main guarantees of the electoral rights and participation of citizens of the Russian Federation in a referendum' have been amended to abolish direct popular voting for the governors, which often served as the grounds for financial-political battles among large corporations.[31]

Despite these efforts, however, scholars of contemporary Russian politics generally agree that the relationship between business and state in Russia remains predominantly informal. Tompson (2005a, pp. 170–1) insists that the state 'has no incentive to abide by any formal commitments to big business and the legitimacy of property rights is still in question'. Hanson and Teague (2005) emphasise the lack of political competition and the absence of clear demarcation lines between the state and the private sector. Roland (2006, p. 97) argues that, despite some healthy growth of the Russian economy, the institutional base 'remains very weak and shows no signs of improvement in the near future'. Yorke (2003), Barnes (2003), and Volkov (2002) claim that regional administrations have been actively involved in the resolution of any major property transfer in their territories and large corporations have often deployed regional institutions to secure industrial assets.

Against the backdrop of these conclusions, however, the first three policy initiatives, which aimed at formalising business-political dialogue in the regions, represented direct borrowings from some well-functioning Western models. Thus, like in the previous chapters we face the problem of the existing gulf between the initial, Western-inspired intentions and rather limited practical outcomes. We shall demonstrate that the implementation of these new policies faced various cultural and contextual constraints. Indeed, Russian political elites executed such borrowed ideas in a particular personalised and authoritarian *style*. This has subverted the original intentions behind these policies and led to diametrically opposite results, effecting a rapidly growing relational informality.

Theoretical and comparative considerations

Before proceeding to elaborate on the nature of cultural constraints, it is important to underscore that the Kremlin's steps towards bringing certain aspects of corporate activity under state control does not make Russia an outlier among the world's leading economies. Looking at the range of Putin's second-term policies towards formalisation of business-political relations in the regions, it becomes clear that these measures do not represent a 'specifically' Russian invention, or a set of regulations aimed exclusively at tightening an authoritarian grip of the Kremlin over the business community. Rather, with the exception of the new gubernatorial elections system, these policies largely resembled measures that have been well entrenched in the West for some considerable period of time.

First, the corporate social responsibility notion emerged in the United States as early as in the 1890s. This was associated with the need to regulate trusts and monopolies, labour-employers relations, and safety in various industries (Farmer and Hogue 1973, p. 2). Comprehensive academic and political debates on the CSR surfaced in the West in the 1950s (Fairbrass 2006, p. 5), and recent decades have seen the proliferation of discussions with numerous seminars, workshops and publications emerging on the market (Cespa and Cestone 2004). New requirements for standardised companies' CSR reporting have emerged. The EU Commission established a European Alliance on CSR, inviting all members of the business community and participating governments to debate on this subject, bring it into the public domain, and render the business-political dialogue in this sphere more transparent.[32]

It is fair to say that the introduction of this concept was not easy. Numerous sceptics (Levitt 1958, Friedman 1970) insisted that maximising profits should be considered as a 'primary social role' of a business. Less pessimistic authors (McWilliams *et al.* 2006, pp. 3–4; Hart 1995; Barney 1991) found that firms should adopt a 'resource-based-view' towards the CSR, thus making socially responsible behaviour more popular among the general public and, thereby, increasing their competitive advantage and profits. Others (Jones 1980; McWilliams *et al.* 2006) contended that the CSR is an ethical issue and, from that point of view, should be a voluntary matter of choice. Irrespective of the multitude of approaches, the literature agrees (see Beesley and Evans 1978, p. 13) that the CSR goes 'beyond compliance and engages in actions that appear to further some social good, beyond the interests of the firm and that which is required by law'.

Second, tax optimisation schemes have been targeted by the governments of most advanced capitalist economies. This is particularly true of transfer price and cross-regional fiscal manoeuvrings. To make my definitions clear, by transfer price I refer to the unit price assigned to goods and services transferred between the parent company and subsidiaries or between divisions within the same firm (Tang 1979, p. 2). If such divisions are located in separate regions with differing fiscal jurisdictions, or even in separate countries, multi-unit corporations could manipulate the level of prices on goods and services transferred between these divisions in order to reduce their tax burden. In particular, multi-unit corporations could artificially reduce the profits of units located in high-tax jurisdictions by undercharging these subdivisions for their supplies, and generate the lion's share of their income within the areas with the most favourable fiscal environment. Subdivisions, in which these companies generate most of their profits, are commonly known as *profit centres* (Plasschaert 1979, p. 24).

The United States was at the forefront of the struggle with transfer price manipulations. The country's fiscal code countering the adverse effects of transfer price manoeuvring is among the most elaborate and comprehensive (Plasschaert 1979, p. 9 and p. 21; Rugman and Eden 1985, p. 247). Australia has implemented a comprehensive regulatory system for transfer pricing since the 1989–90 tax year, and the compliance costs imposed by the Australian Tax Office were regarded as particularly high (Elliot 1997, pp. 25–6 and p. 16). In Japan, transfer price manoeuvring has also been targeted. The June 1975 Business Accounting Principles Board decision, which introduced a series of regulations on transfer price manipulations, affected over 600 companies and left 60 per cent or more of the largest firms with drastically reduced profits (Tang 1979, p. 31). Britain was one of the last industrial nations to introduce regulations on transfer pricing. The 1997 budget provisioned that corporations would not apply the transfer pricing mechanisms for their intra-corporate dealings and rely on self-assessment methods for the transfer price accounting (Elliot 1997, p. 2).

Finally, the impressive array of institutions of business-political collaboration in the West does not need particular elaboration. The multitude of associations for conducting formalised business-state dialogue had begun to appear in West European countries by the end of the 1920s. Such associations were able to replace direct participation of business executives in political institutions

(Cassis 1997, pp. 222–3). The National Recovery Administration in the early part of the Great Depression, the National Resource Planning Board under President Roosevelt, and the Council of Economic Advisors under President Truman serve as just a few examples of the US government formal interactions with business during the first half of the twetieth century (Dimock 1949, p. 7).

Recent decades have seen the emergence of a substantial array of NGOs, discussion groups, seminars, and professional associations aimed specifically at facilitating business-political dialogue. Various Industry (Trade) Associations, Employers' Associations, Chambers of Commerce and Industry have been the most important institutional forms of business-political co-operation since the 1970s. All these organisations have been particularly active in their countries' political lives. The French Movement of Enterprises (MEDEF), the European Confederation of Industries, the Swedish Confederation of Entrepreneurs, the System of Business Chambers in Austria have all at some point delivered their positions to their respective national parliaments, established influential think-tanks, and in some cases organised actions of public protest (Zudin 2005b).

This comparative outlook has direct bearings on the contemporary political situation in Russia's regions. The need to formalise business-state relations at the sub-national level was apparent to political elites both at the centre and in the provinces. More importantly, the policies which have been chosen to implement such tasks have been successfully practised by Western states for many years. This leads to an important question as to why have the results in Russia been somewhat different? We have already argued in the theoretical part of this book that the same measures implemented in differing socio-historic conditions are unlikely to produce identical outcomes. This is due to various cultural and historic constraints. In this light, the political *style* of participant actors has been an important factor in shaping the operational logic of Russia's regional institutions and developmental dynamics of economic reforms. We have also discussed that Russia's unique political style exhibits two most important traits: informality of institutional relations and the tendency towards strong, authoritarian governance. The evolution of business-political dialogue in the regions further confirms these assumptions. Indeed, while with the exception of the new electoral system, the newly adopted policies were not authoritarian in their content, the *style* in which such initiatives have been implemented was personalised and single-handed. The execution of this range of policies in such a manner resulted in the continuation, or even growth, of the informal component in the regional business-political dialogue. This situation once again illustrates the theoretical provisions of Dahl (1956, p. 83, p. 36, and p. 22) who speaks of the impossibility of imposing certain political behaviour by institutional means alone and emphasises the importance of social checks and balances that stem from 'the system of social indoctrination'. The lack of such indoctrination, as well as the commitments of participant actors to operate the new institutional system in a particular democratically oriented manner, once again demonstrates that the process of Russia's regional politics significantly diverges from the intended structures. Let us now look more closely into these four policy areas.

Corporate social responsibility

The idea of corporate social responsibility was introduced in the wake of the *Yukos* affair. This was done partly with the aim of clamping down on big business fiscal manipulations and the 'asset stripping' mentality. During the 1 July 2004 meeting of the Russian Union of Industrialists and Entrepreneurs (the *RSPP*), Putin outlined his vision of corporate social responsibility. He stated that the CSR concerned the adoption of a diligent approach towards fiscal contributions, as well as a more proactive stance towards investments into the sphere of science, education, and human resources.[33]

Russia's political, economic, and academic circles have generally welcomed these ideas. In November 2004, the 14th RSPP Congress adopted 'A Social Charter of Russia's Business'. This document outlined the most important ethical principles for the development of Russia's large corporations and main dimensions of their social mission. These included a general commitment to proliferating the national industrial growth, the development of programmes for the achievement of social peace and solidarity, as well as the provision of human rights and social security (Litovchenko 2004, p. 9). Moreover, following the Western example, a large number of Russian corporations adopted a unified standard for the CSR reporting. In the wake of these efforts, Russia's Federation Council estimated that 200 companies have become engaged in over 350 different CSR initiatives since the beginning of 2005.[34]

At the same time, while the Business Charter clearly outlines that participation in any CSR undertakings should be a voluntary matter of choice – thus warning against coercive policies of the state towards large corporations (Litovchenko 2004, p. 10) – the political vision of Russian officials on this subject was somewhat different. The *way* in which such notions have been implemented has largely been dependent on the existing *political style* of the country's administrative elite.

General adherence to hierarchical and personalised political interaction has been clearly visible. Regional governors have used the initative to rid themselves of the influence of large corporations and begun to actively deploy these ideas with the aim of exerting administrative pressure on enterprises operating in the local economic arenas. This, in turn, has increased the informality in the relationship between the two parties and has provided the opposite effects to those initially expected from the launch of such policies. Marina Moskvina, the general director of the Analytical Centre for the Development of Social Partnership under the Council of Russia's Employers, observed the lack of transparency in business-political collaboration and suggested that most CSR initiatives have been imposed on large corporations 'from above' in a particularly administrative and non-transparent manner.[35]

These claims have been supported by the Association of Russia's Managers, which conducted a comprehensive survey study of this policy understanding by business executive, general public, and regional officials. This report suggests that the regional leaders viewed the CSR thesis in mostly coercive terms. A large number of regional politicians insisted that intimidating big business into implementing various CSR initiatives represents one of the most effective political levers of power. Other officials, in an attempt to move away from the overt relational informality, suggested that a range of legal coercive measures should be adopted in order to enforce compliance.[36] More importantly, such an approach was registered not only among the regional executives but also among various representatives of civil organisations, regional media, and the general public.[37] At the national level, the Federation Council issued an analytical report, which claimed that 'in the relationship between business and state, business plays an explicitly subordinate role'.[38]

This hierarchical and personalised style of implementing the CSR thesis has surfaced across many regions. It is indicative of *how* the former deputy Mayor of Moscow Valerii Shantsev deployed the CSR idea in order to expand the patronage of his institution over the regional business. He notes: 'apart from the existing legislation, there is also social justice. We don't intend to force anybody but, at the same time, we warn: you will not be able to execute any project without our participation. In the end, every defiant entrepreneur will come back to us. But at this point, we will say that we are sorry and that no business had ever been able to succeed without the approval of regional authorities'.[39]

The Kemerovo region was, perhaps, the most illuminating example of this approach. Governor Tuleev forwarded a virtual ultimatum to large industrial corporations with a request to increase salaries. He allowed the firms just one year to consider his 'offer' and bluntly declared: 'I have a clear-cut position: those who do not raise salaries are not welcome in our region!'[40] Tuleev's conflict with the *Amtel Tyre Holding* has proven the seriousness of his intentions. The governor demanded that the company signed an agreement stipulating a 20 per cent wage increase and, in the case of refusal, threatened to raise fiscal duties levied on the enterprise.[41] A range of similar agreements has been put in place with the *Anzherskaya-Yuzhnaya* mine and a number of UGMK subsidiaries, which have been committed to increase wages up to 30 per cent since 2005.[42]

Furthermore, the regional administrations deployed the CSR thesis with the aim of diverting a substantial part of their direct responsibilities to large corporations.[43] Khanty-Mansi Autonomous Okrug, for example, had over 40 differing social programmes in place in 2005, which were implemented with the active participation of large corporations.[44] A 2005 *TNK-BP* (Russo-British oil corporation) press release accounting for the company's CSR commitments nationwide states that the company has altered its policy from implementing selected projects in certain territories to adopting an all-encompassing responsibility for the social sphere within the regions of its industrial coverage. More importantly, the paper pointed out that, in pursuit of these goals, the company would draw upon the 'positive experience of the former Soviet economy'.[45]

Finally, the federal authorities also deployed the CSR thesis in order to place compliant representatives of big business into selected gubernatorial positions. This was done under the assumption that these executives' financial contributions to the regional budgets would promote the development of the social sphere. When the owner of *Sibneft* Roman Abramovich wished to relinquish his duties as a governor of Chukotka Autonomous Okrug in December 2006, the federal authorities 'persuaded' the oligarch to retain his position.[46] Similarly, the plenipotentiary representative to the Siberian Federal District Konstantin Puliko-vskii suggested that Viktor Vekselberg, the owner of *Renova* and a partner in *TNK-BP*, should govern a new Kamchatskii Krai, which was to be formed in 2007 following the merger between Koryakiya and Kamchatka.[47]

Electoral effects

Putin's September 2004 initiative to cancel direct popular voting for the regional executives was partly aimed at checking the rapidly developing informal ties between business and politics in the regions and tackling the spread of the oligarchic monopolism system across the land. Alexandr Veshnyakov, the then head of Russia's Electoral Commission noted, 'given the enormous difficulties that we have faced in fighting corruption during the regional elections, we decided to abolish them altogether'.[48] This was formalised by Federal Law No. 159-FZ of 12 December 2004 and presidential decree No. 1603 of 27 December 2004, which established the system of gubernatorial appointment by regional parliaments on the introduction by the president.[49]

However, the new system has yet again encountered the constraints of Russia's political style and failed to formalise business-state relations in the regions. Despite the fact that large industrial holdings have been deprived of the electoral power instrument, political agents from both sides have utilised the new system to introduce alternative methods of informal, personalised interaction.

First, big business has begun to create informal coalitions to lobby for the support of one or another candidate in the Kremlin during the appointment process. Thus, the selection process has become the subject of informal negotiations within a newly formed political triangle involving the federal centre, the country's main enterprises, and the regional authorities.

Russia's resource-rich and industrial regions, in which economic stakes of large corporations are particularly high, serve as a good illustration. The nomination of the Irkutsk Oblast governor Boris Govorin, who was backed by *Rusal* (Russian Aluminium) and *Alfa* during the July-August 2001 elections,[50] was delayed by the Kremlin for three months. This took place mainly due to the conflicting interests of large corporations. *TNK-BP*, which had stakes in developing the Kovykta gas well, *Interros*, which was determined to explore the *Sukhoi Log* gold mine, *Sual* (Siberian-Ural Aluminium Company), which has four main factories in the region – all took part in ferocious pre-electoral lobbying, backing their own candidates.[51] Due to the lack of a wide agreement among these corporations Putin nominated the head of the Irkutsk *RAO ZhD* branch Alexandr Tishanin to govern the region.[52]

The reappointment of the governor of Khanty-Mansi Autonomous Okrug Alexandr Filipenko, who was particularly skilful in balancing the interests of the region's economic players, represented the result of an informal agreement among *Surgutneftegaz*, *TNK-BP*, *Lukoil*, *Slavneft*, and *Rosneft*.[53] The appointment of Yurii Neelov in Yamalo-Nenetsk Autonomous Okrug, the area of *Gazprom*'s most vital interests, was the subject of negotiations between Russia's President Vladimir Putin and the head of the company Alexei Miller. Neelov had visited both Putin and Miller before the Kremlin officially introduced his candidacy for voting by the regional legislature.[54]

Second, regional business elites have begun to informally influence the choice of regional executives not only during the appointment process but, significantly, during governors' terms in office. Indeed, under Law 159-FZ, the president reserves the right to dismiss the governors at any point during their tenure in office,[55] which could take place as a result of lobbying efforts on the part of regional business. Economic agents have immediately deployed this mechanism so as to make the regional leaders more vulnerable to their political strategies and ambitions. The dismissal of governor Vladimir Loginov of Koryakiya on 9 March 2005 was preceded by a ferocious media campaign launched by anti-gubernatorial circles.[56] It is indicative that the newly appointed governor Oleg Kozhemyako is an owner of Russia's largest fish shipping company *Preobrazhenskaya Baza Tralovogo Flota*, which is registered in neighbouring Kamchatka.[57]

Most gubernatorial dismissals that took place in 2007 occurred due to the governors' inability to conduct a constructive dialogue with local and federal economic elites. The most recent cases concerned the governor of Nenetsk Autonomous Okrug Alexei Barinov, who failed to balance the interests of oil corporations in exploring the Timano-Pechora mines,[58] the governor of the Republic of Khakassiya Aleksandr Lebed, who fell out of Oleg Deripaska's favour,[59] the governor of Sakhalin region Ivan Malakhov, who was unable to establish amicable relations with the region's oil major *Rosneft*.[60]

Third, given that the right of gubernatorial nomination has been transferred to the regional legislatures, these institutions, as well as federal parties, have been placed at the epicentre of regional political life. However, such an institutional arrangement, which is clearly geared towards the formalisation of political activity, has failed to eliminate the informal style with which actors operate the new settings. The increase in political importance of regional assemblies and federal parties has not effected the delegation of political activity to specific agents such as professional politicians, lawyers, and lobbyists. Neither has it led to the emergence of such a class of actors, which is currently absent from Russia's regional landscape.

In turn, these developments fostered *direct participation* of representatives of big and medium-sized firms in the regional legislative process, further consolidating the business-state fusion. While coming to work within regional legislatures and parties, business executives do not break their previous links with corporations and continue to fulfil their executive responsibilities. Russian experts (Chirikova 2005, p. 203) estimate that by 2005, the share of direct representatives of big and medium-sized business in selected regional parliaments had reached some 70–85 per cent.

Regional branches of federal parties have also become attractive targets for political investment by medium-sized and big business. Personal participation in party institutions has become particularly prestigious. Here again, stylistic traits become important. Most executives deployed their partisan and legislative participation mainly as a means of improving the positions of their companies in local economic arenas. Therefore, these businessmen-turned-politicians usually refrained from constructive declarations that would go against the 'general line' of the Kremlin and often failed to attend voting sessions that did not concern issues related to their corporate problems.[61]

The following statement by an owner of a regional retail chain in Yaroslavl is most revealing: 'A seat in the legislature provides me with various administrative privileges. I joined this institution not only due to my political convictions but also with the aim of improving my business positions and obtaining an easy access to the city mayor, whose office is located just next-door. Not every businessman has such a privilege'.[62] Analogous sentiments have been expressed in Saratov Oblast. Olga Alimova, a regional Duma deputy, claims that most parliamentarians viewed seats in the legislature as an administrative safety net for their enterprises and not as a means of socio-political activity.[63] Similarly, a member of the Tver Union of Industrialists and Entrepreneurs Alexei Motorkin openly declared that the influx of big business into the regional legislature had got 'little to do with real politics and is mainly associated with the ambitions of large corporations to resolve various developmental problems related to their enterprises'.[64] The extent of the problem in Perm Oblast led local academics and experts to convene the Conference on 'Perspectives for the Development of Representative Bodies of Power'.[65]

This stylistic pattern has a strong continuity with the Yeltsin era, when the regional assemblies were mostly composed of key members of the former Soviet economic and administrative elites (Lane and Ross 1999; Nicholson 1999). These actors stalled the emergence of a formal approach to politics, in which delegation of political activity to professional agents represents a commonly practiced norm. This also differs from Western Europe, where direct participation of big business in politics had gradually disappeared by the end of the 1920s. Moreover, in Britain, where the parliamentary attractions for big business have been much stronger than in Germany and France, only every fifth chief business executive has at some point been a member of Parliament (Cassis 1997, pp. 222–3).

Finally, the entry of large and medium-sized corporations into the regional branches of *Edinaya Rossiya* and *Spravedlivaya Rossiya* parties is another indication of this stylistic trend. Being the Kremlin's main creations, these parties represent the easiest mechanisms through which business could influence the regional political processes without entering into a direct confrontation with the federal centre. The access of regional executives into the *Edinaya Rossiya* ranks began during the first term of Putin's presidency. For example, the governor of Krasnoyarsk region Khloponin became a member of the *Edinaya Rossiya* Supreme Council with the aim of demonstrating his willingness to collaborate with the

Kremlin on important policy issues.[66] The president of Adygeya Khazret Sovmen, a notable player in Russia's gold market, also developed his political career in conjunction with the party.[67] Similarly, Zelenin, a representative of *Nornikel*, ran in the 2003 gubernatorial election in Tver as a ER candidate.[68] A similar situation occurred during mayoral elections in Magadan in which successful entrepreneur Karpenko, sponsored by a number of corporations, also had the backing of *Edinaya Rossiya*.[69] By 2006, these developments had led to a situation, in which political councils of the regional ER branches have become divided between high-ranking regional bureaucracy and business executives.[70] Similarly, the SR December 2007 regional electoral lists mostly comprised representatives of medium-sized business and owners of regional-scale local enterprises.[71]

Taxation and licensing

The vast majority of Russia's political elite has recognised the apparent need for restructuring in the spheres of taxation and licensing. Due to the growing inter-dependence between business and politics in Russia's regions, favouritism in tax collection had become widespread by the end of Putin's first term. Those companies which managed to secure strong political positions within sub-national administrations obtained a number of official and unofficial benefits that allowed them to contribute just a fraction of their income to regional budgets. At the same time, corporations which had a comparable industrial output but were less successful in establishing political connections within these institutions often contributed a lion's share of regional budgetary income. This had an immediate impact on the provision of social services and welfare in the regions.

Krasnoyarsk Krai is a case in point. The nomination of *Russian Aluminium* executive Nikolai Ashlapov to the post of vice-governor of Krasnoyarsk Krai resulted in a situation in which the fiscal contribution from the company's regional branch *Kraz* in 2001 comprised just 7 per cent of the regional budget's revenue, while *Norilsk Nikel*, which had a similar production capacity, supplied up to 60 per cent of the Krai's financial resources (Tomberg 2002). Due to these apparent problems, which were related to the region's fiscal and economic policies, the quality of welfare provision in Krasnoyarsk – one of the richest and largest regions – rapidly declined. The region became the national leader in wage and pension arrears, with 11 billion roubles of creditors' debt in 2002.[72]

Moreover, nearly all large business holdings employed various sophisticated schemes to reduce their federal and regional tax liabilities. For example, *Sibneft*, which was owned by *Millhouse Capital*, a Russian offshore registered in Britain, developed (mainly by employing disabled people, who have substantial tax exemptions, on its board of directors) a legal scheme to contribute just 10 per cent of its income in 2002, while federally established tax duty stood at 32 per cent.[73] Similarly, *Russian Aluminium* ran up a substantial debt to the budget of Krasnoyarsk Krai during the same year.[74]

Furthermore, the negative economic effects from transfer price manoeuvring – which served as one crucial mechanism of profit maximisation in Russia[75] –

have been widely discussed. The Ministry of Energy claimed that in 2001, the state suffered some \$2 billion loss from the usage of transfer price system in the oil sector alone. With the exception of state-friendly *Surgutneftegaz*, the system has been deployed by all vertically integrated business groups, including *Lukoil, Sibneft, TNK, Rosneft, Slavneft,* and *Yukos* in the oil sector, and *Mechel, Severstal Group,* and *Novo-Lipetsk Metallurgy Plant* in metallurgy.[76]

Transfer price manipulation is closely associated with the usage of tax havens, both domestic and international and, from that point of view, can be very profitable (Rugman and Eden 1985, pp. 283–5; Plasschaert 1979, p. 6). In Russia, one of the most important external offshore schemes deployed by the national oil corporations was associated with the city of Baikonur, Kazakhstan.[77] The system emerged from personal agreements between the Mayor of Baikonur Gennadii Dmitrienko and some of Russia's largest oil corporations. Nine companies – affiliate structures of *Lukoil, Sibneft* and a number of Bashkir enterprises – have registered their head offices in the city and been granted substantial tax benefits.[78] They subsequently rented petroleum factories located in Russia and took responsibility for their fiscal duties. The tax imposed on the production and sale of these products was paid to the Baikonur government at a negligible rate. Meanwhile, these refining plants officially ceased production in Russia and were relieved of their standard fiscal duties, apart from those levied on the profits raised from 'renting' the buildings.[79]

Russia's Account's Chamber estimated that, due to the deployment of this one scheme, the budget lost some 13.8 billion roubles (\$500 million) in 2001 alone.[80] The profits generated by Russia's oil companies via the Baikonur system were so large that these firms engaged in ferocious legal battles with the Ministry of Taxation, which attempted to put the brakes on this system.[81] The matter was subsequently taken to the Russian Constitutional Court, whose 27 June 2005 decision No. 232 has finally recognised the system as illegal.[82]

In addition to the external offshore systems, Russian corporations extensively deployed a range of internal offshore zones. This was coupled with the use of investment benefits, and has led a number of regions into serious financial difficulties. There were three main offshore zones in Russia – Mordoviya, Kalmykiya, and Chukotka – and twenty more regions practicing the investment benefit system.[83] The programme offered substantial tax exemptions for large corporations seeking to invest in economies of Russia's regions and register their enterprises within local fiscal jurisdictions.

In practice, however, the tax benefits generally outweighed the sums that the corporations were required to invest under such programmes. The vast majority of these agreements were concluded with enterprises that had political access to regional administrative institutions. Thus, the companies managed to accumulate tax savings to the order of billions of dollars in a matter of years. At the same time, the regions which practiced this investment system continued to experience budgetary deficits and therefore had to rely extensively on federal transfers.

For example, *Yukos*, through its internal offshore company *Fargoil* registered in Mordoviya, received some 36 billion roubles of tax benefits in exchange for a guaranteed investment of just 1.2 billion roubles in the republican social sphere. As a result of this operation the company increased its income by $1 billion in just two years![84] The state in its politically driven attack on *Yukos* has been able to capitalise on carefully established schemes of tax avoidance practiced by both company executives and accountants.

Minister of Finance Kudrin insisted that in 2002 the state was deprived of revenue of some 42.9 billion roubles due to the internal offshore and fiscal benefit systems.[85] The situation was exacerbated in 2003 when the regions lost 60 billion roubles to these schemes, the cancellation of which was stalled by representatives of big business in the national legislature in 2001.[86] Therefore, it was unsurprising that 60 per cent of all budgetary contributions in 2002 and 2003 was generated by just 10 subjects of the federation and most regional and local budgets were largely comprised of transfers from the federal centre.[87]

At the same time, despite the apparent need to reform such a system, the style with which these reforms were carried out was particularly personalised and informal. With the exception of the abolition of the internal offshore zones in January 2004 and the substantial increase in oil export duties and taxation from 2005,[88] the existing legal framework for transfer pricing mechanisms has not been altered. A broad range of tax optimisation methods that are capable of relieving the companies from the standard 24 per cent of fiscal contribution down to some minuscule 3.5 per cent are still in existence (Radygin 2004). Articles 40 and 20 of Russia's tax code that ultimately outlined the regulations for intra-corporate dealings remained unchanged. In particular, Article 40 stipulates that the existing level of transfer price can deviate from the market price for up to 20 per cent, leaving out the mechanism for determining the fair level of market price.[89] Article 20, on the other hand, outlines how a number of companies could be regarded as 'inter-related' and thus forming the links of the transfer-price chain.[90]

It appears that making an example of the former head of *Yukos* Mikhail Khodorkovskii and his associates,[91] who were the most outspoken advocates of transfer price networks, represented one main mechanism for clamping down on the system.[92] As Tompson (2005b, p. 345) observes, 'quite apart from formal changes in tax legislation, the *Yukos* affair brought about both a change in the informal rules governing oil companies' tax behaviour and an increase in the state's ability to appropriate oil rents directly as a result of its expropriation of *Yukos* assets'.

Indeed, following the launch of the *Yukos* affair, the vast majority of Russia's industrial giants voluntarily declared their refusal to deploy all legally existing transfer-price mechanisms in their inter-corporate dealings. More importantly, a number of firms voluntarily returned funds accumulated from the usage of such a system to the government. The World Bank reports that in 2004 Russia's fiscal authorities collected some 470 billion roubles as voluntary tax debt disclosures, while in 2003 this figure was just 150 billion.[93] Moreover, during the first quarter of 2004 the Ministry of Taxation had already registered a ten-fold increase in the collection of tax on mining.[94] Following the closure of the Khodorkovskii affair in 2005, *Sibneft* increased its declared profits 15 times and returned its fiscal duties to the federal budget at a record 27.5 per cent rate![95]

It is also indicative that *TNK-BP*, a corporation that has strong political protection in the West through its direct relation to the British Petroleum, continued to deploy transfer price mechanisms even in 2005 on the basis of it not being legally outlawed.[96] In 2006, the company enjoyed substantial tax benefits in the Orenburg region, allowing it to contribute at a staggeringly low 13.5 per cent rate.[97] These examples demonstrate the existence of a clear political style. The willingness of Russian corporations not to deploy legally allowed mechanisms due to fear of informal state coercion illuminates the ultimately informal way of conducting politics in Russia.

More importantly, despite the public commitment not to rely on transfer price mechanisms, a large number of companies still deployed various methods of industrial manoeuvring as an important argument in their dialogue with regional authorities. Industrial manoeuvring involves relocation of companies' profit centres and centres of intensive industrial development and modernisation to the regions, where relations with the governors are most amicable (Pappe 2005, p. 89). The ability to engage in industrial manoeuvring is particularly available to multi-divisional corporations, which have a number of equal units located in differing subjects of federation. By generating its main profits in certain territories – without reducing general fiscal commitments to the federal centre – a company could have an important influence on the regional political process. This is particularly true of the regions where industrial intensity is low and the regional budget is dependent upon a handful of large enterprises. It is in this connection that Zubarevich (2005, p. 68) observes that in the metallurgy sector the relations between governors and representatives of big business are generally stronger than those in the oil industry. This is not a surprise, given that industrial diversity in metallurgy regions – such as Vologda or Krasnoyarsk – is less substantial and the ability of industrial manoeuvre for a number of metallurgy companies – such as *Norilsk Nikel*, *NLMK* (Novo-Lipetsk Metallurgy Plant), or *Severstal* – is less impressive.

Nevertheless, industrial manoeuvring represented a particularly worrying factor for a large number of governors. For example, in Sverdlovsk region, the Economic Council called a special committee to estimate which companies, participants in the local market, concentrate their profits in alternative regions.[98] The authorities of the Yaroslavl region have been equally concerned by the fact that a large number of companies register their head offices in Moscow and accumulate most of their profits in the capital, which subsequently benefits from their fiscal contributions.[99] Perm region was particularly concerned with such behaviour on the part of *Lukoil*,[100] while the Irkutsk authorities managed to strike a deal with *SUEK* (Siberian Coal and Energy Company) to concentrate all its profit under regional jurisdiction from 2005.[101]

Furthermore, the system of licensing represented an equally important source of informal interaction between business and politics in Russia's regions. We have already observed that the previous version of the law 'On Natural

Resources' allowed the governors to redistribute mining licensing and agreements in conjunction with the federal centre. The aforementioned 'two keys' principle implied that the regional leaders' signatures were required on all mining licenses.

With such a system in place, investment and privatisation have been greatly affected by corruption. Gubernatorial decision-making has been largely determined by the interests of 'friendly' corporations rather than by objective economic realities. Perhaps the most extraordinary situation arose in Nenetsk Autonomous Okrug in March 2001. Governor Vladimir Butov helped his political partner Andrei Vavilov, the owner of *Severnaya Neft*, to obtain mining rights over the large Val Gamburtseva oil well. Vavilov finally paid just $7 million for the rights in a regional auction, at a time when competitors *Lukoil* and *Sibneft* were offering as much as $100 million![102]

It does not come as a surprise that, during his second term, Putin attempted to alter this system. The 2004 abolition of the 'two keys' rule effectively transferred control over the country's natural resources to the sole jurisdiction of the federal centre.[103] At the same time, despite the introduction of a stricter control over licensing, the informal business-political agreements in this sphere were perpetuated. The recent conflict around Peter Hambro Mining, Russia's second largest gold mining company, serves as an illustration. The firm purchased a number of additional shafts, which would increase the company's market capitalisation, with no practical intentions of mining. Such violation of the mining agreement should have caused license withdrawal. At the same time, regional officials imposed a fine on the company for just 10,000 roubles (approximately US$300.00) in October 2006![104]

Generally, the violation of licensing terms has been widespread. *Gazprom-neft*, for example, breached over 48 mining agreements in 2006.[105] Most oil companies have been usually 35–45 per cent behind their contractual obligation for conducting seismic surveys, 30 per cent behind their contractual promises on exploration drilling, and 22 per cent behind their plans for oil search drilling. Moreover, a large number of companies do not exploit their shaft funds in accordance with the existing licensing code, which allows corporations to leave around 10 per cent of their existing wells undeveloped. *Lukoil*, *Rosneft*, *Gazpromneft*, *Tatneft* have 16–19 per cent of their wells unused. *Sibneft-Chukotka* has developed only 75 per cent of its licensed wells.[106] *TNK-BP* has 30 per cent of all its wells at a standstill, while *Yukos* – as much as 20 per cent.[107]

The Russian Ministry of Natural Resources reports that, since the beginning of 2006, it has withdrawn over 80 licenses in various mining industries.[108] At the same time, such a process has continued to be selective and has often been used by the central state for the achievement of various political goals. This particularly concerned the federal government's disputes with foreign investors such as *BP*, *Shell*, and *Total* over some important mining sites in Kovykta, Sakhalin, and Shtokman. Oleg Mitvol, a senior official within the Ministry of Natural Resources, admitted that the process of license withdrawal is often non-transparent and driven by differing various political considerations.[109]

This stylistic approach continued the line of Putin's first term, when the Kremlin employed a number of administrative measures to expel certain companies – such as *Yukos*-affiliated structures in Yakutiya and in Mordoviya – from some regions. The affair over the Talakansk oil well, for example, exhibited an explicitly political dimension. The state continually manipulated two *Yukos* structures *Sakhaneftegaz* and *Lenaneftegaz*, registered in Mordoviya, on the issue of mining agreements. These had to be revised every twelve months, thus leaving the wells inoperational and keeping companies in the dark vis-à-vis their investment and industrial activities.[110]

It is also important that in 2003, Russia's State Council approved the idea of replacing existing mining licences, which last for at least 20 years, with 3–4 year-long mining agreements that provided the government with further political leverage over big business.[111] In addition to the changes in fiscal and licensing policies, the centre introduced greater political control over the energy sector by altering the composition of main market players to its own advantage. The creation of large projects to explore Eastern Siberia, in which state-owned and state-friendly companies – *Rosneft, Surgutneftegaz, Gazprom*, and *Lukoil* – secured an important role is indicative of this process.

New institutional structures

During the first term of Putin's presidency, the Kremlin established a wide range of institutions for functional co-operation between business and politics at the national level. Among the most important bodies were the Council of Entrepreneurship under the aegis of the Russian Government, the Russian Union of Industrialists and Entrepreneurs, the *Delovaya Rossiya* association, and the Union of Russia's Entrepreneurial Organisations *OPORA*. In addition, the role of the Trade and Industry Chamber has been greatly increased with the appointment of Evgenii Primakov as a head of this institution at the beginning of 2002 (Zudin 2005a, pp. 37–9).

Such developments resulted in a partial institutionalisation of centre-regional economic dialogue. This particularly contrasted with Yeltsin's style of governance. Russia's first president met the so-called oligarchs at ad-hoc meetings only twice throughout his entire career: on 15 September 1997 to negotiate the privatisation of Russia's state telecommunications giant *Svyazinvest*, and on 2 June 1998 to discuss the impact of the world financial crisis on the Russian economy.[112] All other contacts had an unofficial character and many important decisions were taken by the president and representatives of big business behind closed doors. Although Putin had only six official presidential meetings with industrial leaders during the first two years,[113] these became more frequent during 2003 with four meetings in total. Such assemblies were organised by the Russian Union of Industrialists and Entrepreneurs (RSPP) and the Council on Entrepreneurship under the aegis of the government of the Russian Federation. The scope of the questions discussed extended to administrative reform in the regions, corruption among regional and central officials, tax reform, including centre-regional inter-budgetary relations and the system of internal offshores, and developmental issues in Russia's North and in Siberia. The overall importance of these problems for regional affairs testified to the entrenchment of big business in Russia's regional political system.

The development of such co-operation aimed to institutionalise relations between the centre and large business, thus bringing the power resources of the economic elite out of the shadows and making their influence over state policy more transparent. For example, Deripaska – the head of Russia's aluminium giant *Rusal* – planned to effect changes in the established Mergers and Acquisitions Procedure in order to facilitate the company's economic expansion into Russia's regions. During the Yeltsin era, such issues were resolved behind closed doors. In 2002, the tycoon had to at least formally rely on the RSPP to forward an official letter containing his proposals to the Prime Minister.[114] Similarly, the RSPP drafted and forwarded to the government various strategies for improving the investment climate and business environment,[115] for determining the main direction of social policies and insurance,[116] and submitted the 11-point plan for the strategic development of economic policy.[117] All these initiatives became a matter for discussion at the highest echelons of Russia's administrative elite. Moreover, the lobbying drive of the RSPP in the process of drafting a liberal package for currency regulations represented the main reason why this discussion obtained an explicitly public character and had to undergo a number of lengthy hearings in the third State Duma.

The federal assembly also became a key venue in which big business defended its interests. The third State Duma comprised differing factions and inter-faction groups that represented the interests of large corporations. The *Energiya* inter-faction group headed by a direct *Yukos* representative Vladimir Dubov advocated the interests of Russia's main oil companies. The People's Deputy faction headed by Gennadii Raikov represented *Alfa Group*.[118] The Union of Right Forces stood on the positions of *RAO UES*, while the CPRF represented Russia's agrarian lobby. Business representation in the fourth State Duma, elected in December 2003, has been even more substantial. Conservative estimates suggest that approximately 114 out of 450 deputies were business executives. Moreover, 68 people (approximately 60 per cent) have had previous legislative experience, which could make them effective lobbyists (Pappe 2005, p. 83). By the end of his presidency, Putin had decided to prevent Russia's richest executives, in particular those occupying the first 50 places in the Russian Forbes list, from taking part in the December 2007 parliamentary election and deploying parliamentary seats as a means of legal protection and advancing commercial interests.[119] This appeal, however, had little influence. Suleiman Kerimov, the co-owner of *Nafta-Moskva* whose Forbes-estimated fortune comprises some US$12.8 billion, seemed to be the only victim of this new rule. Many of Russia's billionaires, such as Andrei Skotch (US$1.7 billion Forbes estimate), Viktor Rashnikov (US$9.1 billion), Vladimir Gruzdev (US$850 million), as well as a considerable number of large corporations' executives with multi-million dollar fortunes, remained within the *Edinaya Rossiya* party lists.[120]

The abundance of venues for institutional co-operation between business and politics at the national level served as an example for regional authorities. A number of institutions have been established in the regions to resemble the federal experience. Of particular importance were gubernatorial Economic Councils, regional centres for strategic development, industrial and professional unions, and associations of medium-sized and small businesses.

The practice of concluding individual bilateral agreements between regional administrations and selected companies has also gained prominence. Such agreements usually regulated issues related to these corporations' fiscal contributions to the regional budgets, as well as financial commitments towards various social programmes. The vast majority of Russia's leading enterprises, including *Gazprom, TNK-BP, Lukoil, Rusal* and others, have concluded bilateral agreements with regional administrations nationwide (Peregudov 2003, pp. 255–9).

At the same time, as Tompson (2005a, p. 172) observed, 'despite the authorities' attempts to institutionalise forms of state-business interaction ... Russia's commercial elite continues to rely more on personal networks and informal relationship in dealing with state than on formal rules and institutions. All the business clans continue to invest heavily in maintaining networks of favourably disposed politicians and officials at federal, regional, and local levels'.

Such arguments are particularly illuminated by various researches into the extent of corruption and bribery. The Transparency International (TI) index, which annually estimates corruption rating worldwide, placed Russia on the 126th, 127th, and 143rd positions out of the 159 assessed countries in 2005, 2006, and 2007 respectively.[121] These findings were in line with the World Bank and the European Bank for Reconstruction and Development survey that questioned 601 Russian companies. The paper states that, by 2005, bribes and 'kickbacks' had grown substantially in the redistribution of important state and investment orders.[122] Russia's Independent *Indem* Fund also estimates that the volume of bribes paid by large corporations to various officials at all levels amounted to $316 billion per year in 2005. This means that an average company pays $135,800 in bribes annually.[123]

Despite the fact that the situation has marginally improved for medium-sized and small business,[124] their reliance on the informal means of interaction is nonetheless substantial. Polls conducted by the All-Russian Congress of Medium-sized and Small Businesses, held in June 2005 in Yekaterinburg, were most revealing. Every fourth respondent admitted to having bribed regional officials, and only 20 per cent of all interviewed would prefer to rely on institutional networks in order to facilitate their transactions with regional authorities (Nemytykh 2004).

Similarly, polls conducted by the National Project Institute in Perm, Nizhnii Novgorod, and Samara regions demonstrate that only a small number of entrepreneurs were aware of the existence of formal associations for business-political interaction. Four per cent of the respondents in Nizhnii Novgorod were able to name such institutions and, although the figures rose up to 25 and 30 per cent in Perm and Samara Oblasts respectively, most of the respondents insisted that such structures were ineffective.

Furthermore, Gubernatorial Economic Councils have failed to become meaningful venues for business-political co-operation. This is particularly important, given that the Councils have been called to represent the main vehicles of institutional dialogue between the two parties and aimed at resembling the Council on Entrepreneurship created under the aegis of the presidential administration on 5 August 2000.[125] Indeed, both actors failed to operate these Councils in line with the spirit of institutional transparency. A large number of regional politicians insist that 'big business has always deployed alternative financial resources to resolve its main problems individually, working around these structures, which they often considered decorative'.[126] These bodies meet very infrequently, leaving a large room for informal bilateral interaction between regional administrations and large corporations. For example, Tver Economic Council is scheduled to meet twice a year, while a similar body in the Kirov region is supposed to conduct three meetings annually.[127]

Media coverage related to the functioning of these structures is very scarce and existing reports are limited to brief announcements of ongoing meetings. In this context, Oleg Vinogradov, the Yaroslavl regional Duma deputy observed: 'We know very little about how this body functions. What is clear, however, is that all main problems between big business and politics are resolved behind closed doors either bilaterally between the governor and large corporations or with a direct involvement of the presidential administration'.[128]

Finally, bilateral treaties between regional administrations and large corporations, which were intended to formalise all aspects of the dialogue between the two parties, fell prey to similar stylistic characteristics. The governors have begun to deploy these documents with the aim of imposing an informal 'payment' on companies for the right of entry into a regional market. The vast majority of such agreements have been concluded at a time when corporations were attempting to make their first steps in the regional economic arenas. Following the acquisition of the *Nevinnomyskii Azot* factory by the *MDM* group, the government of Stavropol Krai and *Evrokhim* reached an agreement binding the company to implement an extensive package of social development.[129] The conflict between *Inteko-Agro* – an agricultural holding owned by the wife of the Moscow mayor, Elena Baturina – and the Belgorod region administration is also noteworthy. The regional authorities insisted that, in order to be able to operate on the regional market, the company had to finance various developmental projects. In the wake of *Inteko*'s refusal, the regional administration demanded that the federal centre open a criminal investigation into the origins of the company's capital.[130]

The main conclusion that can be drawn from this discussion is that the launch of the new initiatives has not led to the institutionalisation of business-political interaction in the regions. Neither has it placed both actors at the desired 'safe' distance. Rather, interpenetration of business and politics at the regional level took on new forms and dimensions and moved to alternative informal venues.

More importantly, the federal state managed to inflict a certain degree of antagonism between the two parties. The governors deployed the new policies with the aim of exerting direct administrative pressure on large corporations, thus creating, in Bunin's words, a 'serious problem of state dictatorship over business at the regional level'.[131] And while the most explicit forms of informal coercion faded away, the more sophisticated, even somewhat institutionalised, methods of state pressure on large corporations have emerged. Big business, on the other hand, searched for new ways of maintaining and expanding the levers of its informal influence over the regional state institutions. Stakes in forging exclusive relations with the governors grew substantially, as well as the price of direct participation in regional legislative and partisan processes.

Stylistic and cultural predicaments have been the most important factors in shaping the evolution of these dynamics. Indeed, the way in which the Russian elites have understood and implemented the new policies has been somewhat different to that in the West. It also seems that the Russian society has not been entirely ready to embrace certain aspects of social policies borrowed by the government from external sources. At the same time, Welch (1993, p. 164), supported by many other observers such as Geertz (1975), Alexander (2000, pp. 35–6) and Brown (2005, p. 187), notes that cultures should be viewed not as a 'set of givens but as a process' that evolves while shaping and being shaped by institutions. From this point of view, cultural and stylistic changes could emerge, given that the implementation of Western-inspired policies would not bring recurrent socio-political shocks that could make a cultural reversal likely (Alexander 2000, pp. 42–3).

Indeed, some positive shifts have already been registered by a number of observers. Roland (2006, p. 96) notes that public opinion towards big business has become much more optimistic, even among the most conservative strata of pensioners, while Bunin (2005) insists that a new, more sophisticated and institutional, entrepreneurial culture has begun to emerge. However, if the country's political elite does not embrace these attitudes and relevant stylistic approaches wholeheartedly, Russia could invariably face difficulties in implementing any range of institutional practices borrowed from the West. More importantly, given the current informal-authoritarian model of business-political interaction in the regions, the country might face some serious tensions in all fundamental dimensions of its socio-political life, in particular in the relations between business and politics, business and society and, most importantly, state and society.

6   The unintended consequences of gubernatorial appointments in Russia

This chapter will examine one of the most important features of Russia's federal reform, one that was aimed at dramatic alteration of the centre-regional political interaction. The discussion will concern the cancellation of direct popular voting for the governors and its implications for Russia's federal system.

The cancellation of gubernatorial elections was aimed at tightening the central grip over the regions and checking the influence of regional elites at the sub-national level. Arguably, a range of reforms aimed at restructuring primary and auxiliary pillars of the Russian federation, which we have discussed in the three previous chapters, pursued similar goals. However, most of these initatives' structural elements did not contradict the basic principles of the federal theory, and it was mainly the *style* with which such policies have been implemented that raised doubts. The new line towards gubernatorial elections, however, represents an instance of an overt hierarchical, centralising, and almost anti-federal undertaking. However, as Elazar (1987, p. 198) notes 'in federal systems what may seem on the surface to be a centralising activity … may actually have equally strong decentralising tendencies'. In this light, we will continue our discussion of the emerging gap between implemented structures and real political processes in Russia and exemplify Elazar's proposition with the country's regional electoral procedures. We will argue that, despite the appointment of governors undermining the ethical basis for the federal separation of authority, the new institutional settings have resulted in a range of unintended decentralising consequences. In Chapters 3 and 5 we have already observed that this initiative has failed to eliminate the elements of regional pluralism. Here we will develop these themes in greater detail. We will argue that the most significant decentralising effects of this policy include the consolidation of regional elites, the growing significance of regional assemblies, accelerated development of national parties in the regions, and the emergence of highly competitive elections to regional legislatures. The resulting growth of regional, as well as statewide opposition could well prove more problematic for the Kremlin than preserving the traditional electoral system.

Theoretical and comparative considerations

The evident rift between the initial intentions and unintended outcomes represents the most fascinating aspect of the new institutional reform. Two broad themes beg investigation in this context. The first, intentions-related, area questions the extent to which the new policies were geared towards centralisation and how far they violated (or otherwise) the main organisational and moral-ideological principles set out for federations. The second, outcome-oriented, group examines how well institutional mechanisms, chosen by the Kremlin to implement these objectives, were capable of securing the desired results. Were these methods predisposed to aid the Kremlin's intentions or were they to create underlying dangers and transitions leading to unanticipated outcomes?

In the theoretical chapter of this book, we have already argued that the moral and ethical basis for the federal separation of authority implies a substantial degree of independence (and it is only the degree of such independence that is in question) between the central and regional governments. For federalism was set primarily against the hierarchical and, as Elazar (1997, p. 239) notes, 'down a pyramid', devolution of powers and, from this point of view, rejects the explicit subordination of regional authorities to the federal centre. In this context, though not automatically denying the principle of centre-regional separation of powers, inhibits its foremost ethical foundations. For such a system makes regional authorities sub-ordinate (as opposed to co-ordinate) to the state level of government and dependent upon (as opposed to independent of) its confidence and approval.

Comparative experience underpins this point. The nomination of provincial executives is practiced mostly in those federations which have endured an authoritarian regime that circumvent the meaning of federalism. Brazil, which fell prey to military rule from 1964 to 1985 (Samuels 2000, pp. 48–50), Venezuela, which experienced decades of autocratic governance and the military regime between 1948 and 1958 (Griffiths 2002, p. 361), Nigeria, where severe regional disputes led to the introduction of military rule in January 1966 (Elaigwu 2002, p. 75), Pakistan, which has gone through three Martial Laws since its independence of 1947 (Ali 1996, p. i), are cases in point. Even though these countries continued to formally call themselves federations, the ethical basis of such 'federalism' has always been highly dubious.[1]

It would, undoubtedly, be an exaggeration to brand Putin's governance as a military, or even highly authoritarian, regime. However, the introduction of gubernatorial appointments relied on a distinct pattern of the pre-existing unitary and centralising traditions. In the USSR, which was an extremely centralised authoritarian system, republican party secretaries were appointed. Following the dissolution of the Soviet Union, the leadership of the newly formed Russian Federation has been somewhat hesitant to introduce direct popular voting for the governors, and took the opportunity of the 1993 constitutional crisis to ban this practice (Solnick 1998, p. 50). Re-introduction of gubernatorial elections in 1995–6 was swiftly followed by attempts to secure closer central control over the regional leaders. The Anatolii Chubais administration, consulted by Leonid Smirnyagin, Georgii Satarov, and Emil Pain, intended to increase the role of presidential representatives in the regions (Lysenko 2002, p. 40).

With the advent of Putin, central interference in gubernatorial elections accelerated further. During the first term of his presidency, the Kremlin actively participated in all gubernatorial elections taking place at the regional level. For example, the President of Yakutiya Mikhail Nikolaev – who had fair chances of re-election – had to withdraw from the race on 12 December 2001 following direct interference by the Central Electoral Committee.[2] In North Osetiya, the most threatening competitors to acting President Alexandr Dzasokhov fell out of the summer 2001 race due to administrative pressure from the centre.[3] The Krasnoyarsk electoral crisis of autumn 2002 was personally resolved by Putin with the virtual appointment of Alexandr Khloponin to the post of regional leader.[4] Generally, the Kremlin's attitude towards the composition of regional political elite has been highly selective and dependent on a range of subjective factors such as the personality of a regional leader, the political and economic situation within the republic, the existence of natural resources and other politico-economic considerations (Mikheev 2002).

In a number of cases, where the regional leaders were unacceptable for the Kremlin, the centre embarked upon aggressive campaigns for those governors' dismissals. Such a tactic was deployed in Ingushetiya (2001) and Adygeya (2002), where the centre managed to push through its favoured candidates Murat Zyazikov and Khazret Sovmen. In those regions, where compromises were possible, the centre chose the tactics of pressure and intimidation followed by a subsequent inclusion of 'tamed' leaders into the existing power vertical. This took place in Yakutiya (2001), Kalmykiya (2002), and Bashkortostan (2003). Finally, the centre openly supported some of the most amenable leaders during their electoral campaigns and incumbencies in exchange for clearly demonstrable political loyalty. Tatarstan, North Osetiya, and Kabardino-Balkariya could serve as the most illuminating examples of this trend.[5]

In this light, the abolition of direct popular voting for the governors can be viewed as a logical continuation of these centralising arbitrary trends. At the same time, against the backdrop of the evidently undemocratic and anti-federal intentions, the new institutional settings could give rise to a number of unintended, decentralising results. The experience of federal polities which practiced the system of appointment of regional executives demonstrates that these institutional dynamics often result in the consolidation of regional political elites. These could then construct sizable bastions of anti-centre opposition and become potentially menacing to the authority of the general government. In Nigeria, despite 30 years of military rule and central nomination of regional administrations, the regional elites retained a sizeable degree of autonomy and, under the rule of General Gowon, were accused of being too independent for a military regime (Elaigwu 2002, p. 76). The cancellation of gubernatorial elections during the period of military rule in Brazil failed to eliminate state political elites' autonomy and destroy their state-based organisational structures. Therefore, the governors managed to regain their historical influence at the national level during the period of transition, and began to impede federal efforts at democratic reform by acting as 'veto players' (Samuels 2000, pp. 48–50). Similarly, in Pakistan excessive centralisation and appointment of governors, accompanied by authoritarian rule, produced an adverse effect, and has led to the growth of regionalism which posed challenges to the system's stability (Ali 1996, p. 17).

More importantly, the abolition of direct popular voting for the governors represents, in its current form, a move away from a presidential towards a parliamentary regime in the regions.[6] Scholars (Shugart and Carey 1992, pp. 117–24; Verney 1959, pp. 17–56; Shugart and Mainwaring 1997, p. 15) have used a variety of definitional yardsticks for drawing a distinction between presidential and parliamentary systems, as well as for determining criteria for hybrid president-parliamentary and semi-presidential models. However, there is a broad agreement on the two most fundamental criteria. First, in presidential systems, there is a popularly elected executive, while in parliamentary regimes executives are selected by legislatures. Second, while in presidential systems cabinet survival in office is not subject to legislative confidence, and executive term in office is constitutionally fixed, in parliamentary systems the cabinet is responsible to the legislature and can be dismissed from office by a legislative vote of no confidence or censure (Lijphart 1984, p. 117). (The removal of the president in a presidential system is possible in exceptional circumstances of criminal wrongdoing by the process of impeachment.)

A strict classification of Russia's regional systems requires further examination on a case by case basis. What is clear, however, is that institutional regimes in the regions have overall moved much closer to a parliamentary system.[7] Indeed, that the regional parliaments are now responsible for the nomination of the executive and can, at any time, impeach this executive means that cabinet survival in office is subject to parliamentary approval and that the regional executives no longer hold direct electoral mandates.

From a theoretical point of view, these institutional innovations could have far-reaching implications. The existing literature on comparative politics (see Lijphart 1984, pp. 154–7; Shugart and Carey 1992, pp. 226–8 and p. 167; Powell 1982, p. 57; Cox 1987) predicts that parliamentary systems are more likely to give rise to the formation of cohesive and programmatic parties. This effect of parliamentarism could be further reinforced by the proportional representation electoral system which, theoretically, has a strong multiplying effect on the effective number of parliamentary parties and acts as 'an accelerator' for the party-building process.[8] Curiously, Putin introduced this institutional setting a few years earlier. The law 'on the main guarantees of electoral rights and participation of the Russian Federation citizens in a referendum',[9] which was adopted in May 2002, demanded that half of regional assemblies' deputies must be elected through party lists.[10]

We can, therefore, assume that Putin's institutional innovations should, theoretically, result in a growing significance of regional assemblies, increased party competition during regional legislative elections, and accelerated party-building.

The first question that springs to mind in this context is how these developments could influence the dynamics of centre-periphery relations. Would these institutional consequences interfere with Putin's intentions to secure a central grip over the provinces? An analysis of the existing comparative literature affords a conclusion that the Kremlin's centralising plans could be thwarted for at least two reasons.

First, a rapid development of national parties in the regions could give rise to the accommodation of the pre-existing ethno-cultural and socio-political territorial claims. Drawing on a historical experience of Western Europe, Rokkan and Urwin (1982, pp. 435–6) insist that, once surfaced, peripheral sentiments are unlikely to disappear even if their wishes are granted. More importantly, they claim that national political parties cannot remain indifferent to regional demands, and either support the existing sub-national movements or press for cross-territorial integration. The latter option is difficult to implement, once territorial pressures become salient. Thus, statewide parties, in addition to their traditional integrative function, often act as conductors of peripheral demands (Rokkan and Urwin 1982, p. 433). Italy is a case in point. Rapidly developing national parties represented the main force behind regional devolution during the 1960s. Similarly, it was under the aegis of the Christian Democrats, Communists, and the Social Democrats, that the country enshrined directly elected regional government in its 1948 constitution (Putnam 1993, p. 19). It is also significant that regional movements employ the strategies of working upon or within the existing statewide parties. This takes place by either establishing cross-party ties (as was the case in Swiss Jura) or by seeking close alliances with specific national parties (as was the case with the Galician Union de Centro Democratico in Spain, the Swedish People's Party in Finland, the Slesvig Parti in Denmark, and the Italian local parties of Val d'Aosta and Alto Adige) (Rokkan and Urwin 1982, p. 432; Rokkan and Urwin 1983, pp. 148–9).

Second, in addition to the purely peripheral forms of protests the emergence of a viable statewide opposition at the regional level could become another source of concern for the Kremlin. Lessons of comparative federalism demonstrate that the emergence of an effective national opposition is directly dependent on the efficiency of its representation in the regional assemblies and on the political significance of these bodies (Friedrich 1968, pp. 63–5). In Germany, for example, the Social Democratic Party was able to gain national prominence through the progressive achievements in certain *Länder* and its ability to form sustainable coalitions with other parties at the regional level. The Canadian example saw Liberals and Conservatives reinforcing their oppositional stance through effective manoeuvring within the regional legislatures. This was in spite of defeat at the federal level. A similar process took place in India, where the long predominance of the Congress Party has been challenged by the growing complexity and political significance of regional assemblies (Corbridge 1995, pp. 110–11). Moreover, theoretical accounts claim that, in relatively homogenous multi-unit systems – Germany being a prime example – regional elections often enhance the political standings of a statewide opposition by acting as vehicles for the 'punishment' of national ruling parties. This is particularly true at national mid-terms when voters take the chance to express their discontent with central government policies (Jeffery and Hough 2003, pp. 201–2). In heterogeneous multi-unit polities – such as Spain and Canada – regional elections are even more salient, which increases the unpredictability of voting patterns at the national and regional levels (Jeffery and Hough 2003, pp. 209–10).

This brief outlook affords the suggestion that Putin's institutional innovations could facilitate the development of a statewide, as well as a regional, opposition to the Kremlin. This could take various forms. Consolidated regional elites, national parties co-operating with pre-existing regional movements, and nationally oriented oppositional pressures could challenge central attempts at centralisation. In the following subsections I will first discuss the Kremlin's constrictive intentions, and then test whether the introduction of the new system resulted in the emergence of unintended decentralising trends provisioned by these theoretical stipulations.

Constrictive intentions and outcomes

Richard Sakwa (2003, pp. 121–3) observes that Putin, in his quest for reforming centre-periphery relations, was torn between two institutional choices. The first, pluralistic statism, would entail constitutional accommodation of the centre-peripheral division of powers and the unimpeded flow of federal law. The second, compacted statism, is somewhat dubious, and characterised by the 'usage of democratic rhetoric combined with the infringement of the federal separation of powers and imposition of the institutions of a national political community from above'.

The introduction of a gubernatorial appointments system has clearly demonstrated Russia's tendency towards compacted statism. Indeed, law FZ-159 directly challenges the principle of the federal separation of authority by allowing the president to disband regional legislatures if they twice refuse the nomination of a governor or express a vote of no-confidence in the incumbent. The president's right of the subsequent appointment of an interim governor without consulting any popularly elected body further exacerbates this trend. That the ethical basis of such a model is rather tenuous has become clearly evident through the violation of the most basic democratic principles. The lack of transparency and fair competition during the appointment process, as well as favouring the *Edinaya Rossiya* party and appointing candidates that have little affiliation with their regions, emerged as the most immediate problems.

First, Putin's Decree No. 1603 on 'methods of selecting candidates for executive positions of the subjects of federation' states that the presidential envoys must introduce at least two gubernatorial candidates to the presidential administration and that these candidates must be selected on the basis of preliminary consultations with representatives of civil society. Written reports on such consultations must be forwarded to the presidential administration along with the candidates' lists.[11] At the same time, institutional mechanisms behind these provisions were not designed to ensure genuine compliance with these minimal democratic standards. Because the law does not stipulate a clear procedure for consultations between *polpredy* and civil society, presidential representatives often ignore this provision. *Polpredy* draft the lists of candidates independently and introduce the potential nominees to the public as a *fait accompli* after forwarding the documents to the Kremlin. Moreover, such a stance was encouraged at the highest political level. An official from the presidential administration observed: 'We shall make sure that *polpredy* will not release the information on selected candidates. The names of the contenders will only be announced after the decision of the president'.[12] Former presidential envoys to the Volga and Far East Federal Districts – Sergei Kirienko (May 2000 – November 2005) and Konstantin Pulikovskii (May 2000 – November 2005) – were particularly outspoken in following this line. The selection of candidates to gubernatorial positions in the Jewish Autonomous Okrug and in the Saratov and Amur regions was strictly confidential. During his visit to Birobidjan, the capital of the Jewish Autonomous Okrug, Pulikovskii bluntly declared that the public 'does not need to know' the nominees names.[13] Putin's dealing with the Moscow executive had, perhaps, the most arbitrary flavour. Sakwa (2003, p. 133), pointing at the enormous wealth accumulated in Russia's capital, aptly notes that 'whoever controlled Moscow could influence events in the country'. Therefore, the perception of a potential challenge emanating from the administrative team of Russia's most important city drove the Kremlin to dissolve the Moscow administration by appointing Deputy Mayors Valerii Shantsev, Georgii Boos, and Mikhail Men to govern Nizhnii Novgorod, Kaliningrad, and Ivanovo regions respectively.[14] Moreover, gubernatorial appeals to the president for the vote of confidence took place in an atmosphere of secrecy, and in most cases the public learnt about the ongoing nomination process only after the president introduced the incumbent for approval by the regional assembly. It is also important that requests for Putin's confidence normally took place as a result of regional leaders' private meetings with the head of state and were not disclosed to the public.[15] Appeals by the president of Tatarstan Mintimer Shaimiev and the governors of Samara and Kostroma regions Konstantin Titov and Viktor Shershunov are cases in point.[16]

The lack of transparency during the appointment process spills over to the institutional sphere leaving various segments of political elites responsible for important decisions within the regional sphere unaware of current affairs and making their functioning unco-ordinated. The situation in Sverdlovsk region was indicative of this trend. The April 2005 visit by the former head of the presidential administration Dmitrii Medvedev to Magnitogorsk for a meeting with the seven presidential representatives was undisclosed even to the members of the *polpred* office in the Ural Federal District. It later became known that the officials discussed the District's cadre policy and the potential replacement of the governor of Sverdlovsk region Eduard Rossel with local *Edinaya Rossiya* leader Alexei Vorobev.[17] Similarly, members of the Territorial Department of the presidential administration were unaware of the fact that the governor of Tyumen

Sergei Sobyanin forwarded his request for confidence to the president.[18] Putin introduced the Amur incumbent Leonid Korotkov to the regional assembly without informing *polpred* Pulikovskii,[19] while the decision of Kursk governor Evgenii Mikhailov to appeal for presidential confidence was a revelation for the Central Federal District's *polpred* Georgii Poltavchenko.[20] Similarly, the Nizhnii Novgorod media has observed in a light-hearted manner that *polpred* Kirienko was the 'last person' to learn that he 'introduced' Shantsev to govern the region.[21]

Second, the absence of competition during the appointments process is another factor that gives cause for concern. The existence of a viable alternative could make gubernatorial nominations more transparent and democratic and bring essential dynamism into regional political life. However, by allowing the president to introduce only one candidate for the vote by a regional assembly, law No. 159-FZ eliminated competition from the appointment process.[22] Furthermore, despite the fact that indirect competition was provisioned by the 'natural' way of appointment, i.e. through the selection of gubernatorial candidates by the plenipotentiary representatives outlined in Putin's decree No. 1603, this process was non-alternative in practice. In most cases, the *polpredy* nominated bogus contenders to their first choice candidates and made their preference clear before voting in the regional assemblies took place. Chief federal inspectors, local *Edinaya Rossiya* leaders, and heads of regional enterprises normally figured as 'technical' candidates creating artificial competition. The lists of potential candidates often included persons who had no intentions of leaving their current positions or incumbent governors, whose careers in regional politics were clearly over.

Perhaps the most extraordinary situation arose in Saratov Oblast. *Polpred* Kirienko announced that the duties of the incumbent governor Dmitrii Ayatskov would not be extended and offered two alternative candidates for the post of regional leader: the head of Saratov's atomic energy station Pavel Ipatov and the head of the local *Edinaya Rossiya* branch Dmitrii Zelenskii. At the same time, Kirienko clearly indicated his preference for Ipatov, which led Zelenskii to withdraw from the nomination process, thereby depriving the president of any choice.[23] Similarly, the list of candidates for executive office in Yamalo-Nenetsk Autonomous Okrug included incumbent governor Yurii Neelov and the general director of the regional branch of *Gazprom*, *Tyumentransgaz*, Pavel Zavalnyi. A top *Gazprom* executive, Zavalnyi did not intend to leave the company. His colleagues and officials from the *polpred*'s office openly admitted that he was a 'technical candidate' included on the list to create competition.[24] The nomination of Sergei Sokol, a deputy governor of Krasnoyarsk region, to the list of potential candidates in Tyva has also followed this pattern with Sokol having no intentions of leaving his post in Krasnoyarsk.[25]

Therefore, the results of voting in regional assemblies were always predetermined, and local officials often prepared inauguration ceremonies and celebrations prior to the vote. This was particularly true in the Kemerovo, Kostroma, Smolensk regions, Primorsk Krai, and in the Jewish Autonomous Okrug.

Moreover, because such appointments took place on a non-alternative basis, regional legislatures voted almost unanimously for the presidential nominees. This took place in 74 out of 77 regional assemblies under review. Only a few mild exceptions entered this picture. In the February 2005 voting in the Amur Oblast legislature the position of governor Korotkov was rather unstable with seven out of 21 Duma deputies voting against his nomination. Similarly, during the June 2007 vote ten assembly deputies voted against the new nomination of Nikolai Kolesov. Tyva (April 2007) and Krasnoyarsk Krai (June 2007) somewhat followed this trend with 15 out of 132 and eight out of 42 deputies voting against the presidential nomination in the first and second cases respectively. It is also important that in those regions, where the incumbent requested presidential confidence and received the Kremlin's blessing, substantial numbers of deputies were absent from the approval voting. This took place in the Leningrad Oblast (ten deputies missed the hearing), Novosibirsk Oblast (22 deputies were absent), Bryansk Oblast (six deputies ignored the hearing), and Vologda Oblast (seven members of parliament were absent). This could be connected to some silent forms of protest, which never entered a wide political arena.

Moreover, the rules of indirect competition provisioned by the decree No. 1603 do not apply to those governors who decided to appeal to the president for a vote of confidence. In this case, the president does not have any choice of candidates, and must decide whether to introduce the incumbent for approval by the regional assembly within seven days.[26] It is therefore unsurprising that regional leaders have begun to view this provision as a loophole in the legislation that allowed them to extend their term in office, thus avoiding any form of political competition. For example, during 77 gubernatorial appointments that took place between January 2005 and December 2007, 49 regional leaders (which accounts for 64 per cent of all nominations) appealed to the president for a vote of confidence. This particularly concerned those regions where the incumbents' positions were threatened and where the potential of being dismissed by the regional assembly or replaced by a more capable candidate was strong. The appeal for presidential confidence by the governor of Samara Oblast Titov, which was approved by Putin on 25 April 2005, is a case in point. Titov has been involved in ferocious legal battles to remain in office since summer 2004. He could also face serious competition for the nomination: the preliminary list of contenders drafted by *polpred* Kirienko comprised over 40 names, including Titov himself, the chief federal inspector for the Samara region Andrei Kogtev, the region's head prosecutor Alexandr Efremov, and State Duma deputies Vladimir Mokryi and Viktor Kazakov.[27] Titov's unstable position within the regions led him to resign in August 2007. The political position of Leonid Korotkov of Amur region was also weak. His relations with the regional assembly were tense and the list of potential candidates drafted by *polpred* Pulikovskii included Vladimir Gryzlov, a cousin of the speaker of the State Duma Boris Gryzlov, deputy *polpred* Yurii Averyanov, and the chief federal inspector Valerii Voshchevoz.[28] These circumstances contributed to Korotkov's decision to forward a request of confidence to the president. Similarly, Rostislav Turovsky from the Moscow

Centre for Political Technology suggests that if the governor of Chelyabinsk region Petr Sumin did not appeal for presidential confidence, large regional enterprises would lobby their candidates who could be equally capable as regional leaders. The general director of the Magnitogorsk Metallurgy Plant Viktor Rashnikov – known for his friendly relations with the Kremlin – is a case in point.[29] Table 6.1 demonstrates the level of competition during the candidates' selection process and votes of the regional assemblies during the first 15 months of the new system's functioning.

Third, the potential of the governors to effectively liaise with the *Edinaya Rossiya* party represented an important selection criterion for the Kremlin and resulted in further standardisation of the regional political space. Indeed, many governors lost their positions due to their inability to establish productive relationships with the ER elites. The most recent cases include Leonid Korotkov (Amur Oblast),[31] Mikhail Prusak (Novgorod region) (Vyzhutovich 2007), Konstantin Titov (Samara Oblast),[32] and Oleg Chirkunov (Perm Krai).[33] That the Kremlin assessed gubernatorial ability to control political situations within their regions by the means of organising effecting electoral campaigns favouring the *Edinaya Rossiya* party exacerbated this trend. At the same time, such an approach kept many 'politically inconvenient' figures in their positions. This concerns the Yeltsin era governors Eduard Rossel (Sverdlovsk Oblast), Viktor Kress (Tomsk Oblast), Nikolai Fedorov (Republic of Chuvashiya), Murtaza Rakhimov (Republic Bashkortostan), and Mentimer Shaimiev (Republic Tatarstan). Clear outliers among the current elites, these governors still represent an important political force capable of ensuring the adequate levels of electoral support for the Kremlin.

Finally, the Kremlin has often appointed the governors who had very little or no political affiliation with their regions. This process was clearly visible with the appointment of the former Moscow city politicians Valerii Shantsev, Georgii Boos, and Alexander Men to govern the Nizhnii Novgorod, Kaliningrad, and Ivanovo regions respectively. At the same time, these figures were well-known players in the Moscow, as well as the national political arena. A range of additional appointments, such as those in the Saratov, Tula, Irkutsk, Kostroma (2007), Amur oblasts, as well as in the Altai Krai and Republic Buryatiya, saw the arrival of unknown federal technocrats in the position of regional governors. In the Amur region, in particular, the centre went explicitly against the wishes of regional elites by introducing an outsider nominee and thus ignoring a substantial list of candidates suggested by local politicians.[34] These newly appointed figures had no affiliation with local elites and were determined to pursue the will of the federal centre within their jurisdictions at all costs. Against the backdrop of the high reappointment rate (only 26 out of 77 governors appointed between January 2005 and December 2007 were new) it has become clear that the Kremlin adopted a dual strategy of introducing its personal technocratic appointees and perpetuating the process of consolidation of the most established regional elites in selected territories.

Table 6.1 Appointment of regional governors between January 2005 and December 2007[30]

	Date	Region	Appointed Governor	Competitors	Y	N	A	NP
1	4 February 2005	Primorsk Krai	Sergei Darkin	Requested a vote of confidence	36	0	0	3
2	17 February 2005	Tyumen Oblast	Sergei Sobyanin	Requested a vote of confidence	24	1	0	0
3	18 February 2005	Vladimir Oblast	Nikolai Vinogradov	Federal inspector in the region Gennadii Veretennikov	28	3	0	5
4	22 February 2005	Kursk Oblast	Alexandr Mikhailov	Requested a vote of confidence	41	0	0	2
5	24 February 2005	Khanty-Mansi AO	Alexandr Filipenko	Alexandr Suvorov (the mayor of Surgut)	25	0	0	0
6	24 February 2005	Amur Oblast	Leonid Korotkov	Requested a vote of confidence	21	7	1	1
7	25 February 2005	Jewish AO	Nikolai Volkov	Alexandr Vinnikov (the mayor of Birobidjan) and Gennadii Antonov (vice-governor)	15	0	0	0
8	3 March 2005	Evenk AO	Boris Zolotarev	Requested a vote of confidence	18	0	0	4
9	3 March 2005	Saratov Oblast	Pavel Ipatov	The head of regional Atomic Station Pavel Ipatov and the local Edinaya Rossiya leader Dmitrii Zelenskii.	35	0	0	0
10	11 March 2005	Yamalo-Nenetsk AO	Yurii Neelov	The head of Tyumentransgaz Pavel Zavalnyi.	21	0	0	0
11	25 March 2005	Republic Tatarstan	Mintimer Shaimiev	Requested a vote of confidence	89	4	1	6
12	30 March 2005	Tula Oblast	Vyacheslav Dudka	Sergei Kharitonov (chief federal inspector in the region)	37	0	1	7
13	7 April 2005	Koryak AO	Oleg Kozhemyako	Introduced by Putin	12	0	0	0
14	18 April 2005	Chelyabinsk Oblast	Petr Sumin	Requested a vote of confidence	41	0	0	0
15	20 April 2005	Kemerovo Oblast	Aman Tuleev	Requested a vote of confidence	35	0	0	3
16	21 April 2005	Samara Oblast	Konstantin Titov	Requested a vote of confidence	22	0	0	0
17	21 April 2005	Kostroma Oblast	Viktor Shershunov	Requested a vote of confidence	17	3	0	3
18	23 April 2005	Orel Oblast	Egor Stroev	Requested a vote of confidence	46	0	0	4
19	14 May 2005	Penza Oblast	Vasilii Bochkarev	Requested a vote of confidence	42	1	0	0
20	30 May 2005	Lipetsk Oblast	Oleg Korolev	Requested a vote of confidence	34	1	1	2
21	7 June 2005	Republic North Osetiya	Taimuraz Mamsurov	–	62	1	0	0
22	14 June 2005	Rostov Oblast	Vladimir Chub	Requested a vote of confidence	40	1	0	4
23	15 June 2005	Republic Ingushetiya	Murat Zyazikov	Requested a vote of confidence	30	1	0	3
24	15 June 2005	Orenburg Oblast	Alexei Chernyshev	Requested a vote of confidence	40	0	1	5

#	Date	Region	Governor					
25	16 June 2005	Smolensk Oblast	Viktor Maslov	Requested a vote of confidence	39	0	0	9
26	13 July 2005	Tambov Oblast	Oleg Betin	Requested a vote of confidence	49	0	0	1
27	26 July 2005	Kaluga Oblast	Anatolii Artamonov	Requested a vote of confidence	33	3	0	3
28	8 August 2005	Nizhnii Novgorod Oblast	Valerii Shantsev	—	36	0	0	7
29	25 August 2005	Altai Krai	Alexandr Karlin	Yakov Ishutin (the head of the regional forestry ministry)	57	0	1	3
30	26 August 2005	Irkutsk Oblast	Alexandr Tishanin	Valentin Mezhevich (Federation Council Deputy)	42	2	0	1
31	29 August 2005	Republic Chuvashiya	Nikolai Fedorov	Requested a vote of confidence	67	2	0	3
32	15 September 2005	Buryat Autonomous Okrug	Bair Zhamsuev	—	15	0	0	0
33	16 September 2005	Kaliningrad Oblast	Georgii Boos	Yurii Shalimov (deputy governor)	27	2	0	0
34	28 September 2005	Republic Kabardino-Balkariya	Arsen Kanokov	Gennadii Gubin (Prime Minister) Khachim Karmokov (Federation Council deputy), Andrei Panezhev (general director of KabBalkRegionGaz)	105	0	0	5
35	10 October 2005	Perm' Krai	Oleg Chirkunov	Requested a vote of confidence	34	3	0	3
36	21 October 2005	Chukotka Autonomous Okrug	Roman Abramovich	Andrei Gorodilov (first deputy governor) and Mikhail Sobolev – (deputy governor)	11	0	0	0
37	24 October 2005	Republic Kalmykiya	Kirsan Ilyumzhinov	Requested a vote of confidence	22	1	1	0
38	31 October 2005	Stavropol Krai	Alexandr Chernogorov	Requested a vote of confidence	23	0	0	2
39	10 November 2005	Republic Mordoviya	Nikolai Merkushin	Requested a vote of confidence	44	0	1	3
40	21 November 2005	Sverdlovsk Oblast	Eduard Rossel	Requested a vote of confidence	43	0	0	5
41	22 November 2005	Ivanovo Oblast	Mikhail Men'	Valerii Mozhzhukhin (chief federal inspector) and Yurii Smirnov (Federation Council deputy)	32	0	0	2
42	24 November 2005	Tyumen Oblast	Vladimir Yakushev	Sergei Vakhrukov (deputy polpred in the Ural Federal District), Sergei Smetanyuk (deputy governor)	23	0	0	2
43	7 December 2005	Komi Republic	Vladimir Torlopov	Pavel Orda (first deputy governor) and Vasil'ev (Federation Council deputy)	27	0	0	3

continued

Table 6.1 Continued

	Date	Region	Appointed Governor	Competitors	Assembly vote			
					Y	N	A	NP
45	20 February 2006	Republic Dagestan	Mukhu Aliev	Said Amirov (Mayor of Makhachkala) and Saidgusein Magomedov (head of republican treasury)	101	1	0	0
46	3 March 2006	Republic Kareliya	Sergei Katanandov	Nikolai Levin (speaker of the regional parliament) and Pavel Chernov (republican Prime Minister)	53	0	0	0
47	7 August 2006	Nenetsk Autonomous Okrug	Valerii Potapenko	Introduced by Putin	16	0	0	4
48	28 March 2006	Ulyanovsk Oblast	Sergei Morozov	Requested a vote of confidence	21	4	0	3
49	10 October 2006	Republic Bashkortostan	Murtaza Rakhimov	Requested a vote of confidence	119	0	0	0
50	7 December 2006	Republic Yakutiya	Vyacheslav Shtyrov	Requested a vote of confidence	60	3	1	6
51	13 December 2006	Republic Adygeya	Aslancherii Tkhakushinov	Khazret Sovmen (incumbent governor), General Bezhev, Ruslan Khadzhebiekov (speaker of the regional parlrament), Adam Tleush (republican representative in the Federation Council), Aslan Khashir (candidate to the Federation Council)	50	1	0	1
52	20 December 2006	St Petersburg Federal City	Valentina Matvienko	Requested a vote of confidence	40	3	0	4
53	14 February 2007	Murmansk Oblast	Yurii Evdokimov	Requested a vote of confidence	17	5	0	1
54	2 March 2007	Chechen Republic	Ramzan Kadyrov	Shaid Zhamaldaev (head of the Grozny municipal district) Muslim Kuchiev (deputy head of the republican presidential administration)	56	0	0	0
55	10 March 2007	Tomsk Oblast	Viktor Kress	Requested a vote of confidence	34	1	1	6
56	6 April 2007	Republic Tyva	Sholban Kara-ool	Sergei Sokol (deputy governor of Krasnoyarsk Krai), Sherig-ool Oorzhak (incumbent governor) Khonuk-Ool Mongush (speaker of the upper chamber of the local parliament)	132	15	0	11
57	23 April 2007	Krasnodar Krai	Alexandr Tkachev	Requested a vote of confidence	61	1	0	0
58	4 May 2007	Moscow Oblast	Boris Gromov	Requested a vote of confidence	50	0	0	0
59	24 May 2007	Omsk Oblast	Leonid Polezhaev	Requested a vote of confidence	36	6	0	2

No.	Date	Region	Name	Nominated by				
60	30 May 2007	Kamchatskii Krai	Alexei Kuzmitskii	Introduced by Putin	34	1	0	1
61	1 June 2007	Amur Oblast	Nikolai Kolesov	Introduced by Putin	23	10	0	2
62	4 June 2007	Krasnoyarsk Krai	Alexandr Khloponin	Requested a vote of confidence	42	8	0	2
63	15 June 2007	Republic Buryatiya	Vyacheslav Nagovitsyn	Introduced by Putin	57	4	0	3
64	16 June 2007	Belgorod Oblast	Evgenii Savchenko	Requested a vote of confidence	29	0	0	3
65	21 June 2007	Vologda Oblast	Vyacheslav Pozgalev	Requested a vote of confidence	25	1	1	7
66	27 June 2007	Moscow Federal City	Yurii Luzhkov	Requested a vote of confidence	32	3	0	0
67	9 July 2007	Khabarovsk Krai	Viktor Ishaev	Requested a vote of confidence	24	2	0	0
68	9 July 2007	Leningrad Oblast	Valerii Serdyukov	Requested a vote of confidence	39	1	0	10
69	10 July 2007	Tver Oblast	Dmitrii Zelenin	Requested a vote of confidence	28	1	1	3
70	12 July 2007	Novosibirsk Oblast	Viktor Tolokonskii	Requested the vote of confidence	74	3	1	22
71	7 August 2007	Novgorod Oblast	Sergei Mitin	Introduced by Putin	23	0	0	3
72	9 August 2007	Sakhalin Oblast	Alexandr Khoroshavin	Introduced by Putin	20	2	0	6
73	29 August 2007	Samara Oblast	Vladimir Artyakov	Introduced by Putin	39	0	4	7
74	18 October 2007	Bryanks Oblast	Nikolai Denin	Requested the vote of confidence	47	6	0	6
75	25 October 2007	Kostroma Oblast	Igor Slyunyaev	Nikolai Maslov (head of the Alexndr Nevskii Fund social movement)	34	0	0	2
76	24 December 2007	Smolensk Oblast	Sergei Antufiev	Introduced by Putin	44	2	0	2
77	25 December 2007	Yaroslavl Oblast	Sergei Vakhrukov	Oleg Kovalev, Mikhail Babich, Valery Galchenko (regional Duma deputies), Yurii Khardikov (the head of the Northern Moscow municipal district)	44	0	0	4

Consolidation of regional elites

The impact of the reforms on the political situation within the regions backed up our initial proposition on the potential consolidation of regional elites and the consequent growth of oppositional pressures by provincial clans. In order to explore these dynamics it is necessary to ascertain how far the Kremlin was prepared to employ the new system in an attempt to tighten its grip over the provinces and which constraints (or otherwise) the legislation imposed on the elites' ability to exercise power.

First, despite the introduction of the constrictive process of selecting the governors, the centre was unwilling to embark on a full-scale reshuffling of the regional elites and a comprehensive dismantling of the existing balance of centre-regional relations.[35] Therefore, the system of appointment was viewed by the Kremlin as a complementary means of securing further control over the regions – a deterrent mechanism, and not as an instrument of direct and immediate coercion. This was mainly associated with the fear of instability as well as additional struggles for power and property that could follow the arrival of new incumbents.

Indeed, virtually all gubernatorial appointments were made with the aim of preserving political stability at the national level and the elite balance established in the regions during the past decade. Therefore, despite the fact that a number of governors represented clear irritants to the Kremlin, they still received the chance of reappointment. Samara, Kostroma (2005), Kursk, Penza and Orel regions, as well as the republics of Tatarstan, Bashkortostan, and Chuvashiya provide examples of such a policy in Central Russia and Volga Districts. The nomination of governor Boris Zolotarev of Evenkiya in the Siberian Federal District is a particularly interesting example, given the incumbent's connections with *Yukos* and his consistent political support for the company.[36] The appointment of the president of Tatarstan Shaimiev was intended to preserve the political status quo both at the federal and regional levels. Indeed, immediately following his reappointment, Shaimiev declared that, despite the fact that the new system represented a move away from democratic principles, it was introduced to ensure stability and safeguard the country against potential disintegration. He personally pledged to work against such a scenario during his extended tenure in office.[37] Similarly, the purpose of choosing Yurii Neelov and Alexandr Filipenko in Yamalo-Nenetsk and Khanty-Mansi Autonomous Okrugs was to maintain an important balance among the oil and gas enterprises that dominate the financial and political landscapes of these resource-rich regions and play a fundamental role in the nation's economic stability.[38]

The emphasis on maintaining the elite balance encroached upon the effective alteration of power in the regions. Of the 77 gubernatorial appointments that took place between January 2005 and December 2007, only 26 regional leaders were replaced with new candidates. This constitutes a 66 per cent reappointment rate. Four leaders left office for non-political reasons: the presidents of Kabardino-Balkariya and Dagestan Valerii Kokov and Magomedali Magomedov resigned due to a serious illness and advanced age respectively, while the governors of Altai Krai and Kostroma Oblast Mikhail Evdokimov and Viktor Shershunov died in car accidents. Two of these positions were filled with candidates representing previous leaders' power groups: Arsen Kanokov of Kabardino-Balkariya and Mukhu Aliev of Dagestan were appointed following the recommendations of their respective predecessors.[39] This situation brings the reappointment rate higher to some impressive 71 per cent. The lack of cadre turnover was part of Russian regional politics even prior to the enactment of the new system. Sakwa (2003, p. 137) notes that in 1999 in 12 of the 17 contests the incumbent won (71 per cent re-election rate); and of the 46 elections held in 2000 to May 2001 challengers won in only 11 (76 per cent re-election rate).

This pre-existing pattern of poor cadre alteration, combined with the high reappointment rate, resulted in a serious stagnation of personnel in the regions. Indeed, a large number of governors ruled their territories since the early 1990s and some had occupied leading positions since the Soviet period. Governor Stroev of Orel region and the President of Tatarstan Shaimiev have held their positions since the Soviet period and their leadership will be extended to over 20 years following the reappointments. The governor of Sverdlovsk Oblast Eduard Rossel and the governors of Khabarovsk Krai, Jewish and Khanty-Mansi Autonomous Okrugs Viktor Ishaev, Nikolai Volkov, and Alexandr Filipenko have held office since 1990 and 1991.[40] Following reappointments, these governorships will be extended to over 19 years. Table 6.2 demonstrates the lengths of gubernatorial terms in office for selected governor who will have held their positions for over ten years by the end of their current office tenures

Second, the legislative framework supporting the new appointment system impeded the Kremlin's intention to preserve the existing balance of power within the regions. The new arrangement effected important changes in the redistribution of political authority in the provinces, in particular in the power potential of regional governors vis-à-vis the elites. By depriving the heads of regional executives of electoral mandates, the centre made these officials dependent not only on the president but also on the regional power clans. The latter could unite against the governor and remove him from office either by instituting impeachment procedures or by refusing his candidacy during the nomination process. This logic granted competing elites groups further incentives to consolidate their efforts in a struggle for regional power resources. Therefore, such a system intensified political tensions among the rival clans and led to an escalation of conflicts within the regions. This has become particularly clear when the legislative assembly of Altai Krai created a precedent of impeaching governor Mikhail Evdokimov. A famous comedy actor who won the gubernatorial elections by popular vote in April 2004, Evdokimov failed to secure control over the region's financial-political society.[43] The oppositional elites of the neighbouring Altai republic followed this path by initiating an impeachment movement against governor Mikhail Lapshin.[44] The March 2005 dismissal of the governor of Koryak Autonomous Okrug Vladimir Loginov was preceded by a ferocious media campaign launched by anti-gubernatorial circles.[45] The February 2006 appointment of Mukhu Aliev in Dagestan took place against the backdrop of severe struggles among competing elite

Table 6.2 Selected gubernatorial terms in office[41]

	Region	Governor	Date on which first assumed office	Date of appointment	Expiration of term in office	Number of years in office
1	Vladimir Oblast	Nikolai Vinogradov	Elected in 1996	18 February 2005	February 2009	13 years
2	Khanty-Mansi Autonomous Okrug	Alexandr Filipenko	Appointed in 1991	24 February 2005	February 2009	18 years
3	Jewish Autonomous Okrug	Nikolai Volkov	Appointed in 1991	25 February 2005	February 2009	18 years
4	Yamalo-Nenetsk Autonomous Okrug	Yurii Neelov	Appointed in 1994	11 March 2005	March 2010	16 years
5	Republic Tatarstan	Mintimer Shaimiev	September 1989 appointed first Secretary of the Republican CPSU Committee Appointed governor in 1991	25 March 2005	April 2010	21 years
6	Chelyabinsk Oblast	Petr Sumin	Elected in 1996	18 April 2005	April 2009	13 years
8	Orel Oblast	Egor Stroev	Appointed First Secretary of the CPSU oblast committee in 1985; Elected governor since 1993	23 April 2005	April 2009	24 years
10	Kemerovo Oblast	Aman Tuleev	Appointed in 1997	20 April 2005	April 2010	13 years
11	Penza Oblast	Vasilii Bochkarev	Elected in 1998	14 May 2005	May 2009	11 years
12	Lipetsk Oblast	Oleg Korolev	Elected in 1998	30 May 2005	May 2009	11 years
13	Republic Chuvashiya	Nikolai Fedorov	Appointed in December 1993	29 August 2005	August 2009	16 years
14	Buryat Autonomous Okrug*	Bair Zhamsuev	Elected in November 1997	15 September 2005	March 2008	11 years
15	Republic Kalmykiya	Kirsan Ilyumzhinov	Appointed president in 1993	24 October 2005	October 2009	16 years
16	Republic Mordoviya	Nikolai Merkushin	Elected in 1995	10 November 2005	November 2009	14 years
17	Sverdlovsk Oblast	Eduard Rossel	Appointed in 1990	21 November 2005	November 2009	18 years
18	Republic Kareliya	Sergei Katanandov	Elected in May 1998	3 March 2006	March 2010	12 years
19	Khabarovsk Krai	Viktor Ishaev	Appointed in September 1991	9 July 2007	July 2011	20 years
20	Belgorod Region	Evgenii Savchenko	Appointed in October 1993	18 June 2007	June 2011	18 years
21	Omsk Oblast	Leonid Polezhaev	Elected in December 1995	24 May 2007	May 2011	16 years

clans led by the former president Magomedali Magomedov and the Mayor of Makhachkala Said Amirov.[46] In this context, the influence of regional economic elites on these processes has been particularly notable. We have already seen that most governors who faced political difficulties, or lost their positions, have not been able to establish a successful dialogue with regional economic and industrial clans. Thus, in order to survive in their appointed positions the governors have to become integrated in one or another competing elite clan and pursue its economic and political interests within their regions.

Party-building and electoral competition

The transfer to parliamentary settings provided a new logic for regional legislative elections and gave rise to an accelerated party-building. This represented a significant move-away from the pre-existing electoral dynamics. Golosov estimates that the average number of parties competing for the regional legislatures during the 1999–2003 electoral cycle was four.[47] He further claims that, since 1995, there has been a clear trend toward a decrease in the number of parties running for the regional assemblies. Stoner-Weiss (2002, p. 136) supports this claim. Quoting the Central Electoral Commission data, she insists that of the 3,030 deputies elected in 71 regional legislatures between 1995 and 1997, only 336 deputies (10 per cent) were elected from the largest 12 political parties in Russia. Similarly, Independent Institute of Elections data shows that only 3.5 national parties on average entered regional assemblies during the elections which took place in 2003 and in the first half of 2004. In a large number of regions – such as Tatarstan, Kalmykiya, Mordoviya, Ust-Orda – only two parties (mainly *Edinaya Rossiya* and the CPRF) were included in regional legislatures.[48]

This situation has begun to change with the enactment of the new system. While Stoner-Weiss (2002, p. 137) suggests that, during the previous electoral cycles, regional assemblies did not represent a 'particularly big prize for Russian political parties', Zudin, supported by a number of Russia's leading analysts, claims that the introduction of new rules has significantly increased the 'price of entry to the regional assemblies'.[49] Table 6.3 demonstrates that, since the introduction of the new legislation in October 2004, the average number of parties competing for the seats in regional legislatures has grown from 4 to 7.2. Moreover, the enactment of the new system coincided with an increase in the average number of parties composing the regional legislatures from 3.5 to 4.5.[50] Interestingly, this number is almost close to the average number of parties in multi-party systems in established European democracies (Powell 1982, p. 80). It is, of course, too early to draw any optimistic conclusions. For as we have already observed, many such organisations, instead of delivering mediating functions between business, public, and state, relied on *direct* participation of top business executives in their structures. Similar worries have been expressed about the consistency of these parties' memberships and the principles of political expediency exercised by many of the participants while entering such organisations. Nevertheless, the trend in itself is interesting and noteworthy.

Table 6.3 Competition during elections to regional legislatures since introduction of new law, January 2005–December 2007[51]

	Region	Date	Number of mandates in regional assemblies provisioned for party lists	Number of parties and blocks taking part in the elections	Number of parties composing legislature
1	Tula Oblast	3 October 2004	24	11	7
2	Republic Marii El	10 October 2004	26	5	5
3	Irkutsk Oblast	10 October 2004	23	9	6
4	Sakhalin Oblast	10 October 2004	15	11	6
5	Chita Oblast	24 October 2004	21	5	4
6	Kaluga Oblast	15 November 2004	20	6	5
7	Kurgan Oblast	28 November 2004	17	8	6
8	Bryansk Oblast	5 December 2004	30	8	5
9	Arkhangelsk Oblast	19 December 2004	31	9	6
10	Republic Khakassiya	26 December 2004	38	8	6
11	Taimyr Autonomous Okrug	23 January 2005	7	4	4
12	Nenetsk Autonomous Okrug	6 February 2005	10	7	5
13	Vladimir Oblast	20 March 2005	19	7	5
14	Voronezh Oblast	20 March 2005	28	13	5
15	Ryazan Oblast	20 March 2005	18	7	7
16	Amur Oblast	27 March 2005	18	9	8
17	Yamalo-Nenetsk Autonomous Okrug	27 March 2005	11	4	3
18	Magadan Oblast	22 May 2005	13	5	4
19	Belgorod Oblast	19 October 2005	18	7	4
20	Republic Buryatiya	30 October 2005	9	3	3
21	Ivanovo Oblast	4 December 2005	24	8	6
22	Kostroma Oblast	4 December 2005	18	8	6
23	Novosibirsk Oblast	11 December 2005	49	7	4
24	Tver Oblast	18 December 2005	17	8	6
25	Tambov Oblast	18 December 2005	25	11	3
26	Moscow Federal City	25 December 2005	20	11	3
27	Chukotka Autonomous Okrug	25 December 2005	6	2	2
28	Taimyr Autonomous Okrug	25 December 2005	30	7	4
29	Chelyabinsk Oblast	25 December 2005	30	7	4
30	Kirov Oblast	12 March 2006	27	9	5
31	Republic Adygeya	12 March 2006	27	8	4
32	Kaliningrad Oblast	12 March 2006	20	7	5
33	Kursk Oblast	12 March 2006	23	9	3
34	Nizhnii Novgorod Oblast	12 March 2006	25	6	4
35	Orenburg Oblast	12 March 2006	24	8	5

Table 6.3 Continued

	Region	Date	Number of mandates in regional assemblies provisioned for party lists	Number of parties and blocks taking part in the elections	Number of parties composing legislature
36	Khanty-Mansi Autonomous Okrug	12 March 2006	14	5	4
37	Republic Altai	12 March 2006	21	13	6
38	Jewish Autonomous Okrug	8 October 2006	8	6	3
39	Sverdlovsk Oblast	8 October 2006	14	10	4
40	Novgorod Oblast	8 October 2006	14	9	4
41	Lipetsk Oblast	8 October 2006	28	8	4
42	Astrakhan Oblast	8 October 2006	29	10	4
43	Primorsk Krai	8 October 2006	20	10	4
44	Republic Chuvashiya	8 October 2006	22	5	3
45	Republic Tyva	8 October 2006	16	6	2
46	Republic Kareliya	8 October 2006	25	7	5
47	Perm Krai	3 December 2006	30	8	5
49	St Petersburg Federal City	11 March 2007	50	6	4
50	Tyumen Oblast	11 March 2007	17	6	4
51	Tomsk Oblast	11 March 2007	21	8	5
52	Samara Oblast	11 March 2007	25	7	6
53	Pskov Oblast	11 March 2007	20	7	4
54	Orel Oblast	11 March 2007	25	8	4
55	Omsk Oblast	11 March 2007	22	7	2
56	Murmansk Oblast	11 March 2007	16	6	4
56	Moscow Oblast	11 March 2007	50	7	3
57	Leningrad Oblast	11 March 2007	25	6	4
58	Vologda Oblast	11 March 2007	17	5	5
59	Stavropol Krai	11 March 2007	25	5	5
60	Komi Republic	11 March 2007	15	7	5
61	Republic Dagestan	11 March 2007	72	6	5
62	Krasnoyarsk Krai	15 April 2007	26	7	5
63	Smolensk Oblast	2 December 2007	24	4	4
64	Saratov Oblast	2 December 2007	18	5	3
65	Penza Oblast	2 December 2007	13	4	3
66	Krasnodar Krai	2 December 2007	35	4	3
67	Kamchatka Krai	2 December 2007	27	4	4
68	Republic Udmurtiya	2 December 2007	50	5	4
69	Republic North Osetiya	2 December 2007	35	9	3
70	Republic Mordoviya	2 December 2007	24	4	2
71	Republic Buryatiya	2 December 2007	33	4	4

Table 6.3 also reflects some empirical evidence that electoral campaigns to the regional legislatures have become extremely competitive and virtually turned into battlegrounds among various movements and independent candidates supported by different groups of provincial elites. For example, the March 2005 elections to the regional assemblies of Irkutsk, Voronezh, Sakhalin Oblasts and Altai Krai revealed the intensity of political competition between *Rodina* and *Edinaya Rossiya*.[52] The March 2005 elections to Ryazan and Vladimir regional legislatures took place with the personal participation of the main national party leaders – Dmitrii Rogozin, Vladimir Zhirinovskii, Grigorii Yavlinskii, Gennadii Zyuganov, Valerii Gartung, and Sergei Mironov.[53]

Moreover, we have already mentioned that the formation of new electoral coalitions has become a distinctive trend. Rostislav Turovsky and Alexei Titkov (2005) from the Moscow Carnegie Centre observed that liberal parties have begun to display left-wing populist tendencies by forming unusual unions with patriotic and social-democratic forces to obtain seats within regional assemblies. For example, *Yabloko* worked with Sergei Mironov's *Party of Life* during legislative elections in Amur Region and Taimyr Autonomous Okrug. This secured the block a clear 18 per cent majority vote in the first case and 22 per cent of the vote in the second. Similarly, the *Union of Right Forces* (SPS) formed a coalition with the *People's Party* in Irkutsk Oblast, thereby managing to enter the regional assembly. Growing importance of regional elections has also prompted closer collaboration between irreconcilable liberal parties SPS and *Yabloko*. The two parties united their efforts during the October 2005 elections to the Moscow City Soviet and Chelyabinsk regional legislature.[54] Similarly, the parties refrained from simultaneous participation in the same regions during the 12 March 2006 campaign that covered eight federation subjects. Russia's liberals took part in seven campaigns with *Yabloko* running in four and SPS in three separate territories.[55] Some candidates had to change party allegiance specifically to enter the race. In Orenburg and Nizhnii Novgorod, a large number of the *Yabloko* candidates were, in fact, acting SPS members who participated in the election under the *Yabloko* banner.[56]

Independent efforts by political parties were also notable. The SPS congress, for example, ruled to 'concentrate political efforts on campaigning in the regions, in particular during the elections to regional legislative bodies'.[57] This gave the party positive results in Kurgan, Amur, Tula, Bryansk, Ryazan, and Moscow Federal City regions.[58] *Rodina* has also become determined to struggle for influence at the regional level and fought hard for every province, employing differing electoral strategies and appealing to various legal and judicial institutions at the national level. These efforts were rewarded. The party managed to overcome *Edinaya Rossiya* during the elections to the Sakhalin regional legislature, entered the regional Duma of Tula Oblast, and gained a third place in Irkutsk and Kaluga Oblasts.[59] Its political potential was clearly evident during the 12 March 2006 campaign in Altai Krai. Following deregistration in the seven remaining regions, the party challenged *Edinaya Rossiya* and gained 10 per cent of the votes.[60] The CPRF was steadily lagging behind *Edinaya Rossiya* and struggled with *Rodina* and the *Party of Pensioners* for the left-wing electoral space. Nevertheless, no regional parliament was formed without its participation. In May 2005, the *Party of Pensioners* – a marginal organisation at the national level – has displayed the most impressive results in the regions by gaining the mandates in the assemblies of Amur, Tula, Irkutsk, Kurgan, Nizhnii Novgorod, Magadan, Kirov, Kalinigrad Oblasts, the republic of Khakassiya, and Khanty-Mansi Autonomous Okrug. In Magadan, the party was competing against *Edinaya Rossiya* for the leadership and secured the second largest number of mandates in the regional assembly.[61] A similar level of success was achieved in the March 2006 Nizhnii Novgorod legislative election.[62] It is also clear that the party's success stemmed from its brand and programme, and not from any significant political popularity of its leaders. The party electoral performance during the 12 March 2006 elections was not hampered by the December 2005 expulsion of its founding father Valerii Gartung.[63] The party subsequently merged with *Rodina* and the Party of Life to form a nation-wide left-wing project *Spravedlivaya Rossiya*.

This trend in party growth was repeated towards the end of the Putin presidency. During the March 2007 regional legislative campaign, CPRF gained seven second and five third places; LDPR entered nine regional assemblies, and SPS managed to succeed in five regions.[64] This led to a situation in which a visible predominance of the pro-Kremlin *Edinaya Rossiya* party at the federal level has been seriously questioned on the regional landscape.[65] The subsequent introduction of *Spravedlivaya Rossiya*, the second party of power, has not helped the situation and increased regional party competition. Within these conditions the Kremlin adopted a range of corrective measures with the view of keeping a grip on party developments and ironing out the emerging political diversity within Russia's regions. By the beginning of 2008, the centre had started to deregister a large number of parties and prevent them from running regional electoral campaigns. This concerned not only the existing oppositional parties but also some loyal pro-Kremlin formations. In Altai Krai, for example, the Agrarian party was refused registration in March 2008, despite its support for Dmitry Medvedev as Russia's presidential candidate. Similarly, the pro-Putin Party of Peace and Unity was denied registration in the Ulyanovsk region during the same time.[66] Furthermore, the Kremlin's general trend towards creating larger national parties by the means of introducing a 7 per cent electoral barrier and stricter administrative procedures had begun to take effect by the end of 2007. By then, only four parties – *Edinaya Rossiya*, *Spravedlivaya Rossiya*, LDPR, and CPRF – had normally taken part in the regional legislative campaigns.

Implications on centre–periphery relations

The rapid development of a national party system in the regions, the growing significance of regional assemblies and legislative elections, as well as the consolidation of regional elites, have demonstrated the potential to undermine the centre's ability to control the provinces. In this case, the decentralist vector of Russia's federal relations has become prominent.

First, the elevated role of the regional assemblies prompted the local deputies to emphasise the significance of their institutions at the national level. Early attempts to defy the centre and reject the idea of gubernatorial appointments were a manifestation of such a trend. Duma deputies in St. Petersburg, for example, adopted this stance to underscore the 'political importance of Russia's second largest city'.[67] Similarly, the leader of Yaroslavl SPS branch Sergei Sushkov notes that 'local deputies were determined to enter national politics and their refusal to adopt the new law was one means of achieving such a goal'.[68] The regional assembly of the republic of Tatarstan also refused to abolish the old version of the law on the election of the republican president and the document was 'temporarily suspended'.[69]

Scarcely less important remains the influence of the new system on the development of executive-legislative relations in the regions. During the Yeltsin era, regional assemblies often represented amorphous pro-gubernatorial majorities unable to effectively communicate with the regional executive (Badovskii and Shutov 1995; Nicholson 1999). The new powers granted to the legislatures enabled the sub-national parliaments to become more assertive in their dialogue with the regional leaders, whose legitimacy mandates have now become subject to parliamentary confidence. Indeed, a number of Russian experts, in particular those actively engaged in fieldwork (Chirikova 2005, p. 204), have already observed the emergence of tectonic changes in the evolution of inter-branch dialogue. Executive-legislative conflicts have become far more frequent with legislatures overturning a large share of gubernatorial bills and proposals. The legislators of Irkutsk Oblast, for example, refused to approve governor Tishanin's protégé for the post of vice-governor and returned the budget drafted by the administration.[70] A similar situation took place in the Yaroslavl and Nizhnii Novgorod regions (Chirikova 2005). This trend is indicative in that the regional leaders represent presidential nominees, and the ability of the regional legislatures to challenge these officials indirectly implies their ability to challenge the federal centre.

Second, the growing significance of regional legislative elections facilitated further trends towards the consolidation of a regional, as well as statewide, opposition. The Independent Political Foundation *Indem* reports that, during the October 2004–March 2005 round of sub-national legislative elections, *Edinaya Rossiya*'s performance fell substantially. In Koryakiya, the party received just 22.7 per cent of the vote, as opposed to the 48 per cent gained during the national campaign (which comprises just 47 per cent of the national result). In Arkhangelsk, the decline was from 37.9 per cent gained during the national race down to 23.6 per cent in the regional race (62.2 per cent of the national result). In Sakhalin, the party received just 17.8 per cent of the vote instead of the national 30 per cent (59.3 per cent of the national result), and in Khakassiya, 23 per cent instead of the national 30.7 (74.9 per cent of the national result) (Kynev 2005). During the 12 March 2006 legislative campaign *Edinaya Rossiya*'s performance was also unimpressive. Turovsky notes that, despite the apparent victory in all eight regions, the party was not able to form majorities in many parliaments.

This was particularly true in Republic Altai and Kirov region, where *Edinaya Rossiya* polled substantially below 30 per cent.[71]

These dynamics largely conform to the theoretical proposition discussed at the beginning of this chapter that the electorate often deploys regional legislative elections as a means of 'punishing' ruling parties. However, the extent of *Edinaya Rossiya*'s regional underperformance seems worrisome in a comparative light. The German data, collected between 1949 and 1990 (Jeffrey and Hough 2003, p. 202), demonstrate that, on average, governing parties polled 88.4 per cent of the vote share that they should have 'expected' on the basis of their national performance. The figures above give reason to believe that the oppositional sentiment in Russia's regions remains particularly strong.

It is also important that many national parties have begun to deploy regional politics as an arena from which they attempt to challenge the Kremlin. During the Nizhnii Novgorod campaign the CPRF organised a number of protests against the communal service reform. These demonstrations attracted thousands of people and resulted in clashes with the police.[72] The appointment of governor Chirkunov in Perm Krai was also accompanied by protests and demonstrations organised by the local CPRF branch, the Agrarian party, and Russia's workers party, as well as various ecological and military movements.[73] Analogous protests were organised by oppositional parties against the reappointment of governor Rossel in Sverdlovsk region.[74]

Equally valid is Urwin and Rokkan's observation on the potential co-operation between statewide parties and various ethno-cultural peripheral movements. In Sakhalin, *Rodina* co-operated with the regionally based *Nasha Rodina Sakhalin i Kurily* movement, while in Tula it worked simultaneously with two regional organisations *Glas Naroda Za Rodinu* and *Rodina – Zasechennyi Rubezh – Partiya Rodina* (Kynev 2004). *Yabloko* entered into an alliance with the Bryansk regional movement *Edinenie* to form a new regional organisation *Za Vozrozhdenie Bryanshchiny*, roughly translated as *For the Revival of Bryansk Region*.[75] SPS, in co-operation with local 'nationalists', created a Sakhalin regional patriotic organisation *The Union of Sakhalin Patriots* (Kynev 2005) while, in Arkhangelsk, the party worked united with the local patriotic movement *Our Motherland – Arkhangelsk Region* (Kynev 2005). It is also significant that such territorially based organisations and movements co-operated mostly with opposition parties, which indirectly implies that socio-political protest in Russia could take territorial dimensions.

Finally, the appointment system altered the means by which intra-regional elite conflicts unfolded and elevated these disputes to the national level – thus creating serious tensions in the centre-regional dimension. This was primarily due to the ability of sub-national parliaments to force the president to introduce a new governor within seven days following an impeachment and to pressure the head of state to refuse the incumbent a vote of confidence.[76] Such an arrangement could make the Kremlin a hostage to the lobbying strategies of regional elites, in particular if the latter came to dominate legislative assemblies.

In many regions this triggered the proliferation of oppositional forces, which began to appeal to the centre to resolve intra-regional disputes. In Saratov region, for example, 19 out of 35 Duma deputies forwarded a note to Putin stating that the assembly would refuse to support the nomination of the incumbent governor Dmitrii Ayatskov.[77] Ayatskov was subsequently removed from office and appointed as Russia's envoy to Belorussia.[78] A similar situation took place in the Nizhnii Novgorod region, where 36 local assembly deputies united against governor Gennadii Khodyrev and forwarded a letter of complaint to the then *polpred* Kirienko.[79] Largely due to this pressure, the Kremlin removed Khodyrev from office and posted Moscow Deputy Mayor Shantsev to govern the region.[80] Ivanovo Oblast, where the Duma deputies declared open political warfare against governor Vladimir Tikhonov, is another case in point.[81] In Altai Krai *polpred* Kvashnin did not include the incumbent Mikhail Lapshin in the list of potential candidates due to severe pressure from the regional assembly.[82] Orel and Kursk regions have undergone similar experiences.[83] The opposition in Bashkortostan also appealed to the Kremlin. Demonstrations sponsored by republican financial groups struggling against president Rakhimov's clan for the redistribution of power and property took place in Moscow, and a note of protest supported by as many as 107,000 signatures was delivered to the presidential administration.[84] In Kamchatka Krai the regional deputies adamantly refused to appoint Oleg Kozhemyako, the governor of Koryak Autonomous Okrug, to the position of governor. This resulted in Vladimir Putin offering a new compromise figure of Alexei Kuzmitskii in May 2007.[85] In the Tyva region, the victory of the *Spravedlivaya Rossiya* party in the October 2006 regional election paralysed the functioning of the local parliament. The Kremlin had to give in to the demands of the regional elites to replace the long standing regional governor Sherig-ool Oorzhak with a more appropriate candidate that came from a winning political clan.[86] Clearly, this makes the federal system less resilient to oppositional pressures, as negotiating directly with hostile elites is far more intricate than employing effective political agitation during electoral campaigns and thus ensuring tenure in office for loyal candidates.

These regional trends could mark a significant shift in the country's political evolution. For they mainly indicate that decentralist tendencies, despite their current adaptation to the rigid institutional realities, are clearly surfacing. The potential for the emergence of a viable opposition undoubtedly represents a positive development for Russian politics, increasingly lacking pluralism and transparent competition. However, it will come as bad news for the current occupants of the Kremlin, who might face the risk of losing their grip over important segments of political society. At the moment, the Kremlin has the ability and sufficiency of political resources to tackle any unanticipated developments in the regional arena. This situation, however, could swiftly change with an alteration of environmental conditions triggered by a potential economic slowdown or unexpected international shocks. It will then become clear that the nature and political stance of the emerging regional opposition, as well as the Kremlin's potential responses to these dynamics, will shape the future of the political process in Russia in its centre-regional dimension.

7 Inter-regional relations and territorial problems

In this chapter we will tackle the problems of inter-regional territorial integration and central attempts at reconsolidating Russia's economic and ethno-federal space. The establishment of institutions of inter-regional functional co-operation in key spheres of politics and economics is an important factor for the development of integrated federalism at the 'vertical level'. The tasks of inter-regional integration in Russia present a wide range of challenges. Of particular importance are the problems of harmonising the existing economic space, creating stronger links of institutional communication within the newly established Federal Districts, ironing out regional disparities within the ethno-federal dimension, and eliminating the most pressing problems of the politico-economic detachment of Russia's most remote regions. The Kremlin attempted to address this spectrum of problems by launching a variety of policy responses in all given dimensions. This ultimately provided regional politics with some integrative vectors. At the same time, the arbitrary stylistic pattern under which these policies were carried out defied the principles of federalism and resulted in the ultimate strengthening of the authority of the central state over inter-regional interactions. This impeded the development of independent regional initiatives and the establishment of a constructive inter-regional dialogue based on the most important socio-economic problems. It is also significant that the centre ignored ethnic and national sentiments while pursuing the inter-regional integration within the ethno-federal dimension. In addition, the Kremlin's selective attitude towards differing regions resulted in the growing territorial disparities and hampered the territorial integration process. We will first discuss the Kremlin's success in the sphere of inter-regional integration and then move on to the analysis of problems pertaining to Russia's territorial and ethno-federal dilemmas.

Central attempts at inter-regional integration and its tentative successes

The Kremlin's attempts at promoting inter-regional socio-economic and political integration have brought a range of positive outcomes. Initially, many analysts questioned the ability of Putin's federal reforms to encourage the development of inter-regional relations. There were two main themes under discussion. First, some analysts suggested that the reforms represent a move toward disintegration of the country into seven separate units. This view was clearly expressed by Boris Berezovskii in his open letter to the president in spring 2000.[1] The second group has insisted that new federal relations would promote inter-regional rivalry with an arbitrary outcome for the centre. As Solnick (2000) writes, 'Putin's early moves suggest a long-range strategy based on pitting provincial leaders against each other rather than provoking them to unite against the centre... the president would use inter-regional rivalries to erode privileges granted to specific regions while simultaneously undermining the unity of regional leaders. As long as regional leaders are focused on eliminating their neighbours' special privileges, the centre can reassert its authority by playing the pivotal role of an 'objective arbiter'. According to Solnick, the centre, in an attempt to hold power, would embark on fostering inter-regional rivalries, disrupting inter-regional collaboration, and preventing governors from adopting unified strategies.

The implementation of the first view was quite impossible due to the developing links between the political activity of the federal centre, its representatives in the regions, and the heads of the regional executives. The second view also seems dubious in that the centre was more interested in achieving a harmonious integration of Russia's regional space that would avoid any sharp conflicts and prevent instability.

From a legal point of view, presidential decree No. 849 – that instituted federal districts and introduced the post of the plenipotentiary representative in these new formations – indicates that the resolution of inter-regional conflicts does not fall into the legitimate sphere of presidential representatives' authority. The document explicitly states that 'representatives shall interfere in the resolution of conflicts between the centre and regions under its jurisdiction', and there is no mention of the ability of the representatives to arbitrate in inter-regional conflicts.[2] This legislation in itself precludes the presidential representatives' ability to 'pit' the governors against each other and makes regular interference in various conflicts unlikely from the legal point of view. In addition, the policy of promoting inter-regional rivalries would only exacerbate the existing relations between the centre and the regions by increasing the number of 'disaffected' (and therefore oppositionist) territories and by destabilising the general political situation on the regional dimension. Such a scenario lies outside the scope of central interests.

Furthermore, in contrast to Solnick's suggestion that inter-regional conflicts would have a political (precisely legal-constitutional) orientation, it seems that with levelling legislation to all-federal standards, there were virtually no conflicts of a political nature among the regions with the exception of the ongoing regional mergers. Rather, disputes took place mainly over financial matters such as investment, industry, property sharing, and taxes. It is not surprising that those adjacent areas which 'share' a number of properties without sufficient distinction as to who is entitled to collect revenues engaged in the highest number of conflicts. Moscow Federal City and Moscow Oblast are entrenched in disagreement over the distribution of income tax. Regional officials claim that residents working in the federal city should pay taxes to the regional budget and not to Moscow city budget, as is the case at present. This situation impedes the implementation of some transportation projects that could increase mobility between the regional population and the federal city.[3] A similar situation occurred in relation to Sheremetevo international airport, which is located on the territory of Moscow Oblast and brings no revenue to the region.[4] Similar disputes surfaced in 2002–3 between Krasnoyarsk Krai and Taimyr Autonomous Okrug over taxation policy for Norilsk Metallurgy Company.[5] Tyumen Oblast, Khanty-Mansi and Yamalo-Nenetsk Autonomous Okrugs were engaged in various conflicts of a similar nature relating to the redistribution of wealth and taxation among these resource-rich subjects.[6] Most regions resolve these conflicts via various settlement commissions, which either operate on a regular basis or represent ad-hoc structures, most importantly without resorting to the assistance of *polpredy*.[7]

It is also important that the central policy towards the role of big business in politics has played its role in the process of inter-regional integration. We have already argued in Chapter 5 of this volume that the centre prompted the competitive expansion of large corporations into Russia's regions. Apart from the wide range of consequences discussed, this process has also had a profound impact on the dynamics of inter-regional integration. The arrival of big business in the sub-national political arena has substantially altered the balance of power within the regions. Large corporations have begun to influence most important political processes taking place within Russia's provinces, including the development of inter-regional political and economic dialogue.

Indeed, when control over economic resources was in the hands of regional leaders, they often employed it to manipulate the centre politically in their struggle for further autonomy. Very often, these claims employed nationalistic rhetoric to disguise underlying politico-economic ambitions. The advance of Russia's industrial groupings into the regions demonstrated that the political agenda of big business differed from that of the governors. While regional leaders pressed for legal-political autonomy within their own territories, the heads of large corporations continued to be aware that their influence would not be confined to just one region. Therefore, the idea of potential expansion to other federation subjects invalidated the incentive to press for political autonomy and legal asymmetry of one particular territory. This vector of regional politics was important in the establishment of uniform rules for inter-regional and centre-regional dialogues – areas in which representatives of big business attempted to effect institutional changes.

For example, in December 2002, the Russian Union of Industrialists and Entrepreneurs established an arbitration committee with a mandate to resolve various corporate conflicts and ongoing oligarchic wars in the regions. Although the decisions of this institution were not legally binding, the committee's real power could rival that of an official court: failure to comply with its rulings could incur dents in executives' reputations and, in the worst case, lead to a boycott by the business community.[8] Because oligarchic wars often took place across different regions and represented a prominent feature of regional political life during the first term of Putin's presidency, the establishment of this institution could be seen as creating closer links of inter-regional co-operation and strengthening federal integration. Similarly, the *Association of Russia's Bankers* organised a special investment fund, which was devoted to the development of legal projects reflecting the problems of Russia's nascent financial community both in the regions and at the centre. The Fund also intended to advocate its legal proposals in governmental structures and in the State Duma. According to Andrei Nechaev, the head of directors of *Rosinbank*, the establishment of this body was a step toward the creation of institutions of 'civilised lobbying' that would assist the development of important legal projects reflecting the common interests of various facets of Russia's business community.[9]

In the economic sphere, Turovsky (2002) observed the emergence of greater industrial and political inter-dependence between the regions and the establishment of a nascent system of checks and balances in their dialogue. Turovsky (2002) associates this with the integrity of industrial cycles. For example, the aluminium industry requires substantial supplies of electricity, which links the interests of the *Russian Aluminium* and *TNK* (*Sual*) with *RAO UES*. At the same time, electricity producers depend on *Gazprom* and the coal-mining industry, which is controlled by various economic groupings in Kemerovo and Irkutsk Oblasts and in the Komi Republic.

Generally, the search for co-operation and institutionalisation of common rules by large industrial groups had clearly surfaced by 2002 – the year hallmarked by a successful series of mergers, acquisitions, and the relatively peaceful resolution of inter-regional financial conflicts.[10] The strategic alliance between *TNK* and *Sibneft* to purchase state oil giant *Slavneft*, the formation of an investment coalition between four leading producers – *Lukoil*, *Yukos*, *TNK*, and *Sibneft* – on a large oil transportation project in Murmansk, and the proposed merger between *Silovye Mashiny* and *OMZ* to create the largest national machinery producer spoke for the existence of integrating economic tendencies in Russia's regions.[11]

Another argument against the federal centre's intentions to ignite inter-regional tensions lies in the fact that Putin's team aimed to assume the leading role in institutionalisation of inter-regional co-operation that had already been underway. Indeed, important developments in this sphere emerged as early as two years prior to the reform and had risen to prominence by the time of Putin's election in March 2000. Inter-regional integration trends had surfaced by 1998, and experts were convinced that these co-operative patterns were likely to grow especially when regional elites come to adopt some form of unified policy. A number of sociologists indicated the emergence of vertical elite co-operation groups and inter-regional or horizontal associations for economic collaboration. The first case implies politico-economic groups comprising influential politicians from Moscow and representatives of regional elites, for example the alliance between regional elites of Tyumen Oblast, *Gazprom*, and Prime Minister Viktor Chernomyrdin, as well as the union between *Mintopenergo* and the parliamentary deputy group *Rossiiskie Regiony*. The second case assumes nine major inter-regional economic associations – such as *Chernozemye*, Siberian Agreement, Great Volga etc. – as well as an association of the Far East governors, the Union of Russian Cities, and the Union of mayors of small and medium-sized cities, which collectively attempted to deliver their proposals to the federal authorities (Chirikova and Lapina 1999, pp. 160–5). Logically, against the backdrop of these developments, the centre had no choice other than to take the lead in this process and to become an important figure behind regional integration. However, a number of analysts such as Vladimir Klimanov (2001, p. 129), Cameron Ross (2002, pp. 63–5) and others argued that the role of inter-regional associations would markedly diminish with the introduction of new federal districts. There were also suggestions that Putin's reform would become a major impediment to the process of development of all forms of inter-regional co-operation.

Arguments for a decline in inter-regional co-operation could potentially extend only to the already existing inter-regional associations, which practically gave way to the Kremlin's sanctioned organisations. We have already observed these trends while discussing the role of the centre in the creation of supportive pillars of the integrated federal structure. At the same time, while the reform indeed altered the dynamic and means of inter-regional co-operation, it did not halt the process itself. Putin's team of politicians opted for assuming the lead in these developments and his innovations intended to position federal districts as templates for collective regional activity in the international arena and in the sphere of inter-regional collaboration. This argument is supported by the fact that one of the main official purposes for creating the federal districts was the promotion of closer inter-regional ties. This was enshrined in the presidential decree 'on plenipotentiary representatives', which established the development of inter-regional co-operation as one of the presidential envoys' direct responsibilities.

The envoys themselves confirmed their commitments to this task. Former *polpred* Kirienko noted in 2002 that, 'while working on the creation of the federal "power vertical" during the first year, representatives plan to devote their second year to restoring the federal horizontal'.[12] Former *polpred* Cherkesov positively evaluated attempts by the federal centre 'to formulate a single policy that would transform previously spontaneous inter-regional co-operation on to an institutional basis'.[13] For the governors, the promotion of inter-regional ties also seemed to be one of the main functions of the federal districts – even their *raison d'être*. Deputy Governor of Moscow region Nikolai Rebchenko noted that the 'creation of federal districts took place purely because of the central government's intention to foster inter-regional ties'. He further argued that 'it is unlikely that inter-regional economic co-operation could be raised to the contemporary level without central initiative', and that 'federal districts will have created a foundation for the future development of inter-regional economic co-operation, which regions with equal potential need'.[14]

The Central Federal District (CFD) displayed rather impressive results in developing the network of inter-regional co-operative institutions, which could even serve as a model for other districts, in particular in the economic sphere.

At the same time, precisely because the majority of these institutions concentrated on the development of the District's economy, *polpred* Poltavchenko managed to establish an unequivocal lead over these processes. This enabled the centre to act as a major player on the 'co-operative field' ultimately controlling the activities of the newly established organisations. In achieving such results, however, Poltavchenko had an important comparative advantage over his colleagues in other districts: the CFD accounted for nearly a quarter of the entire national investment potential and the District needed to modernise over 80 per cent of its existing industrial capacity in order to digest and channel such an influx of funds into its resource-poor territories.[15] The regional authorities were incapable of resolving the modernisation issue with their own efforts, therefore, reliance on central subsidies, persistent control over their redistribution, and the establishment of a single think tank for attracting external hi-tech investment became the most essential components for the District's economic survival. All these circumstances conveniently placed Poltavchenko at the centre of the District's economic and strategic development and enabled him to lead the process of centre-regional functional co-operation. Nevertheless, the practical effectiveness of the newly created institutions was clearly visible.

The Council of the Heads of the Subjects of the Federation in the CFD is one example of such co-operative institutions. This administrative organ is the regional version of the federal State Council, instituted by the president to promote communication with the heads of the regions.[16] It is indicative that the former head of the Federation Council Egor Stroev was nominated to head this body: such a step would enable the *polpred* to gather all the political heavyweights of the District and provide them with certain incentives to work for the benefit of the District and within the District's institutional structures. As Stroev noted immediately following his nomination, 'we shall work hard for the benefit of the Central Federal District – the native core of Russia'.[17] The Council meets no less than twice a month and a number of decisions taken by this body proceed to the federal government or the president for further discussions. Senior federal officials from various ministries, deputies of the State Duma, and members of the Federation Council often attend the Council's hearings. The meetings are also open to representatives of the District's big and medium-sized businesses. Naturally, the most common problems discussed at the hearings concern vectors of economic development of the District, the improvement of infrastructure, communications, and the growth of the agricultural sector.[18] The Council had already developed a system of investment into the District's real economy sector via the issuance of special regional bonds which are released and guaranteed by the local budgets on the condition of higher income from taxation and development. Moreover, the Council helped the District's regions to develop a single strategy in their relations with bordering countries such as Moldova, Ukraine, and Belorussia.[19] The Council also initiated the establishment of the CFD Investment Agency – an institution that plays an important role in the provision of consultations and sponsorship with particular reference to agriculture, the development of high technologies and in the spheres of building and construction

(Kichedzhi 2003). Practical results from the functioning of this institution have already been quite visible. In 2002, the Agency in co-operation with *Vneshek-onombank* established a Russian Leasing Company (*RLK*) to supply and lease agricultural machinery to the farms of Central Russia. With an investment potential of 1.5 billion roubles *RLK* concluded a number of important supply contracts with the Anglo-American Agricultural Company AGCO. Moreover, the CFD's Investment Agency established a number of important building projects: 37,000 square metres of hotel space will be built in the main cities of Russia's historic golden ring. This will be complemented by the construction of an additional 2,774,200 square metres of accommodation in Yaroslavl, Kostroma, Kursk, Orel, Ivanovo, and Tambov Oblasts. As the District's industrial leader, Moscow Federal City acts as the main contractor for these projects.[20] All these developments led to a situation in which the Central Federal District achieved first place in the country for the volume and growth of the building sector. Furthermore, Poltavchenko took the lead in fostering Moscow's functional co-operation with other regions of the District. He noted that 'apart from the restoration of inter-regional economic ties, it was our impressive achievement to turn Moscow economically and politically towards the regions of Central Russia. Today Russia's capital city is capable of investing in the economies of other regions'.[21] Indeed, the Mayor of Moscow Luzhkov established a non-commercial organisation called the Centre for Assistance of Socio-Economic Development of the CFD Regions. Under the aegis of this institution Moscow Federal city and four other donor-regions of the District assumed responsibility for directing funds towards regional investment projects, the fiscal return from which could generate no less than 0.5 per cent of a region's budget (Kichedzhi 2003).

To continue the inter-regional co-operative momentum in the social sphere, Poltavchenko established a number of expert councils responsible for the development of certain areas of social welfare and for enhancing co-operation among regional and federal officials in these fields. The Political Expert Council is one such organisation that attempts to provide assistance in the areas of inter-confessional relations, ethnic problems, maternity and childhood, the development of physical culture and sport, and various youth programmes.[22]

The presidential representatives of other federal districts have also made a lavish number of attempts to initiate and lead integration processes. Leonid Drachevskii, head of the Siberian federal district, was the first to have established a similar District State Council, an institution that was to develop industrial and economic strategies for the district's regions. A consultative body that meets once a month, the Council is comprised of *polpred*, governors of the district's regions, and heads of regional assemblies.[23] Similar Councils were instituted in all other districts. Moreover, *polpred* and the governors of the Siberian federal district acted as a strong regional formation, creating a precedent with the conclusion of an international economic agreement with Mongolia for the creation of transportation links in 2001.[24] A year later, *polpred* Drachevskii conducted an international meeting with the President of Kyrgyzstan Akaev, which once again underscored the *polpred*'s intention to promote the district's position as an inter-regional association capable of establishing economic ties with other territories.[25] The Ural Federal District established direct economic links with the British Council in Russia, while leaders of the Siberian and Far East Federal Districts organised a number of official events promoting economic co-operation with China. The Southern Federal District established an inter-regional balancing committee to settle the economic debts of the district's large enterprises,[26] and a consultative legislative committee promoting inter-regional contacts between regional MPs and parliamentary analytical groups.[27]

At the same time, previously functioning inter-regional associations have begun to work under the aegis of the administrations of federal districts. *Siberian Agreement* and *Great Volga* adapted best to the new federal arrangement. The authorities in the Siberian Federal District headed by presidential envoy Drachevskii established intensive co-operation with existing association *Siberian Agreement*. This co-operation extended to spheres of international and inter-regional affairs and touched upon a number of important economic issues. For example, *Siberian Agreement*, authorities of Siberian Federal Districts and representatives of Mongolia conducted a number of consultative meetings regarding the development of plans of economic co-operation between Siberian regions and Mongolia, which then served as a basis for a visit of the Russian Prime Minister to the country.[28] *Siberian Agreement* took part in organising an international seminar on problems of inter-regional integration and the development of federal districts – a forum that hosted representatives of the Council of Europe, Ministry of Federal Affairs, scientists from the Siberian Branch of RAN and representatives of the *polpred*'s office. Moreover, representatives of *Siberian Agreement* and the *polpred*'s officials jointly drafted a strategy for 'Energy Development of the Siberian Federal District up to 2020' – an important document which assumed the construction of atomic plants in the District.[29] The Volga Federal District developed intensive co-operation among trade unions in the district's regions, *polpred*'s offices, and the *Great Volga* association. These institutions jointly established an association of trade unions of the Volga Federal District's regions. All the regions concluded tri-lateral agreements between employers, regional administrations and the new Association.[30] The new institution conducted a number of political actions related to wage arrears in the District and issued important guidelines for co-operation with the presidential representative in the District.

These examples demonstrate that the establishment of the federal districts and their close co-operation with previously existing inter-regional associations granted some of the latter prominence in the economic and political spheres of inter-regional life. At the same time, it is important to remember that independent functioning of such organisations has effectively come to an end and these movements have been transformed into the vehicles of conducting central policy in the inter-regional dimension.

Territorial problems and challenges

Despite the positive efforts at inter-regional integration, Russia preserved a range of important territorial problems that have yet to be effectively resolved. More importantly, the Kremlin continued to deal with such issues ultimately relying on its traditional style of political expediency, favouritism, and informality. The absence of transparency in centre-regional relations resulted in significant regional differences and impeded comprehensive territorial integration. On the economic front, central selectiveness and favouritism have led to profound socio-economic disparities among the regions. In this light, Russia's economic growth exhibits a dire picture when examined from the regional angle. The Ministry of Economic Development states that in 2007 budgetary sufficiency of Russia's ten poorest subjects composed some 19.8 per cent of the national average, while budgetary sufficiency of the ten richest regions reached as much as 200 per cent of the national average.[31] Similarly, the level of direct investments per head differs 150 times among the regions. The Moscow Institute for Regional Development claims that over one half of regional investments are redistributed among 20 per cent of rich donor-regions, while 20 per cent of recipients attract just 3 per cent of the current investment flows. In this light, the macroeconomic indicators of Russia's richest regions compete with those of the Netherlands and Canada, while the country's poorest territories remain at the level of Kenya and Nepal.[32]

The Ministry of Economic Development raised further concerns by declaring that in 2007 Russia's 12 leading regions were responsible for more than one half of the national GDP. The remaining subjects survived on the federal subsidies, which constituted up to 80 per cent of these regions' budgets.[33] Moscow Federal City stands alone as a particularly good example of Russia's territorial disparities. The city is responsible for 22 per cent of the country's domestic product and, together with the oil-rich Tyumen region, accounts for over one third of the entire regional production (Zubarevich 2007).

It does not come as a surprise that the economic nature of Moscow's detachment from the rest of the country is also translated into socio-political and cultural realms. In 2004, 74 per cent of Russians insisted that Moscow and other regions represented 'two politically incompatible poles'. In 2006, this figure decreased to 63 per cent, still retaining the impressive gap between the capital and the rest of the country. In 2005, only 17 per cent of Russians believed there to be a harmony of interests between the centre and the regions. This figure had improved by 2006 and reached 25 per cent.[34] At the same time, 30 per cent of Russians declared their negative attitude towards Moscow in 2006, and as many as 67 per cent admitted that they would not like to change their place of residence, despite the evidently more prosperous conditions of the country's capital.[35]

Inter-regional tensions have also surfaced within the fiscal sphere. Many regional leaders lamented the situation in which they had to contribute a lion's share of their income to poorer territories in the course of economic redistribution process. These sentiments were mainly expressed through the regional aspirations for an increase in financial independence and political autonomy. Economic concerns relate primarily to excessive political centralisation and redistribution of financial resources in favour of the federal centre. In addition, many regions protested against the current donor-recipient system, in which richer, hard-working, and more successful territories contribute large shares of their income to sustain poorer and less motivated regions.

Some governors sarcastically observed that obtaining a recipient status has become financially convenient. Mayor of Moscow Yurii Luzhkov often branded the current system of fiscal redistribution as ineffective and unconstitutional. Regional elites of the oil-rich Tyumen Oblast support this view. Many claim that the simple redistribution of funds in favour of depressed regions does not improve their macro-economic parameters and living standards.[36] The Mayor of Nizhnii Novgorod Vadim Bulavinov declared that the current system of centre-regional fiscal relations is characteristic of the 1980s USSR and, due to its inefficiency, cannot be sustained in the face of contemporary economic challenges.[37] The President of Tatarstan Mintimer Shaimiev also suggested a revision of the existing centre-regional financial relations by granting the regions greater autonomy in financial-economic, social, and political spheres. In 2007, he particularly lamented the fact that the centre had extended its control towards the spheres constitutionally designated for regional jurisdictions.[38]

These deficiencies have been largely transferred into the socio-political dimension. In this light, Russia has begun to exhibit the emergence of the two significant division lines that cut through the country's regional space. The South, as well as Eastern Siberia/Far East has become virtually detached from the rest of Russia's mainland. This created the premise for political instability, separatism, and economic depression. In the Southern region, the most pressing problems relate to terrorism, national tensions, inter-ethnic conflicts, proliferation of political clan relations, popular protests against local governments, and accelerating unemployment.[39] Putin's government publicly admitted its inability to alter the situation within the region, defeat corruption, and popularise the 'Moscow-centric' mode of thinking among local elites.[40] The centre still continues to appoint these regions' leaders in accordance with traditional clan relations. Altering this appointment policy may further destabilise inter-ethnic relations within these territories and prompt the leaders of existing ethno-economic clans to popularise the ideas of secession from the Russian Federation.[41] At the same time, perpetuation of clan politics alienates the general public and civil society from the regional decision-making process and leads to an increase in public protests that often involve street violence. Such a situation also gives rise to a rapid spread of radical Islamic ideas, which begin to act as a substitute ideology to a genuine public participation in local politics.

Inter-ethnic and religious violence has become widespread in Ingushetiya. In September 2007 terrorist attacks, ethnic crimes, and actions of public protest took place on a daily basis. A number of Russian families were shot, federal journalists kidnapped, and military attacks on representatives of legal enforcement agencies carried out. It comes as no surprise that the local population actively protested against the spread of violence and rapidly declining living standards.[42] Dagestan and North Osetiya repeat this picture with terrorist attacks and ethnic conflicts continuing to claim numerous lives. The former presidential envoy to the region Dmitry Kozak admitted that these negative dynamics stem from the lack of centre-regional, as well as inter-regional integration and the resulting spread of corruption within the regional institutions.[43] Kozak often went as far as to suggest the introduction of a direct presidential rule in a number of key regions.[44]

With regard to Western Siberia and Far East, primary problems relate to the transportation difficulties, expanding migration from China, the growing geopolitical interest of other nations in the region, simmering separatism, and the struggle for resources. While separatist tendencies in the area are currently low, some national territories still harbour such sentiments. The intensity of these could increase given the appropriate socio-economic and political circumstances. Republic Yakutiya represents one such federation subject, albeit other ethnic regions could also follow its lead. In 2006, 54 per cent of the republican population had no trust in the federal government, and 29 per cent of citizens trusted the State Duma against 35 per cent of those who did not.[45] More importantly, republican elites established various political organisations promoting the interests of the region at the federal centre. One such movement, Yakutiya-ALROSA, is backed by as many as 80 per cent of the regional population, while the local Civic Front organisation relies on 46 per cent of popular support. In these conditions, 61 per cent of the republican population favour the introduction of a special status for Yakutiya. Moreover, Yakutiya's 'independence day', when republican sovereignty was declared on 27 September 1990, still remains a public holiday, despite the elimination of the relevant constitutional clause from the regional statues. It is also important that Yakutiya represents the last federation subject to complete the legal harmonisation process. The most contested issues stemmed from Articles 1 and 5 of the republican constitution that declared republican sovereignty and unilateral access to natural resources.[46]

On the economic front, regional domestic product of the Far East is Russia's lowest and the industrial pattern is predominantly resource-based. Regional transportation links are almost absent with no effective means of connecting this territory with the federal centre. An average air ticket between Far East and Moscow, or any other territory of Russia's 'mainland', is priced around 50,000 roubles (US$2,000), while similar long-haul flights offered on the international market within the conditions of commercial competition are priced around US$500–600.[47] It comes as no surprise that the share of transportation costs within the final price of products manufactured within the region accounts for anything from 55 to 70 per cent (with the Russian average standing at 25 per cent).

Such a situation results in a deterioration of the demographic situation within the region. Population decrease is 3.9 times greater than the country's average with Russia's State Statistics Committee insisting that by 2016, the area will host just 6.5 million people.[48] It is also important that young people migrate into the central part of the country, accelerating the current demographic crisis. The problem is exacerbated by the increased migration from neighbouring China. The newcomers are reluctant to integrate into the Russian socio-cultural environment and establish ethnic enclaves that are often oriented towards economic co-operation with their homeland.

It is important that the government has managed to assess the threats resulting from the detachment of the vast pieces of territory from the federal centre. Thus, some important investment programmes have been developed to rectify this situation.[49] In early 2007 Putin instituted a special governmental committee headed by the then Prime Minister Fradkov. The committee was tasked with developing main policy lines aimed at the reintegration of the West Siberian and Far Eastern space with the rest of the Russian territory. The body included a number of regional governors as well as the leading representatives of the Russian Academy of Science. The ensuing federal programme for the development of the Far East until 2013 provisioned an impressive 566 billion roubles (US$22 billion) for various socio-economic initiatives within the entire area. At the same time, there were some limitations to the mode in which such initiatives have been implemented. Despite that these developmental vectors and problematic points were chosen correctly, such projects had an overwhelmingly political character and aimed at enhancing the country's prestige in the international arena. For example, 147.5 billion roubles (US$6 billion) alone were earmarked for the reconstruction of Vladivostok city as a centre of economic co-operation with the Asia-Pacific countries. Given that the city was nominated to host the 2012 Asia-Pacific Economic Co-operation Summit, a large number of four and five star hotels with over 11,000 accommodation spaces will be built. Further, 7,000 rooms of hotel accommodation will be constructed in the Russkiy Island, with additional conferences and VIP facilities.[50]

However, given the dire economic conditions of the Far East area, the post-summit use of these facilities remains unclear. Governmental proposals to transform the city into an international gambling and recreation centre generate more questions than answers. Natalya Zubarevich (2007) from the Moscow State University insists that, instead of channelling such substantial funds towards international projects, the government should have divided the grants equally between Vladivostok and Khabarovsk. The latter, being the second most important city of the Far East, would contribute to the formation of an important infrastructural circuit capable of sustaining a complex development of this territory for domestic needs. In a similar ideologically charged fashion the government suggested a range of grandiose projects unsubstantiated by any economic criteria. Among those were the restoration of the Northern Sea Path, construction of the railroad to Magadan, and the establishment of new industrial complexes in Yakutiya (Zubarevich 2007).

In the case of Russia's South, a development programme was instituted in 2007. The initiative earmarked some 52 billion roubles (US$2 billion) for the regional investment purposes between 2008–12. The document emphasised the development of the transportation infrastructure, the establishment of purpose-built industrial zones, the revival of the regional agricultural sector, and accelerating the tourist industry growth. The programme intended to increase the regional average wage threefold and reduce unemployment from 24 to 19 per cent by 2012.[51] Dmitrii Kozak, the former presidential representative to the Southern Federal District, hoped that such initiatives could transform the regions of South Russia into the fastest growing territories in the country. At the same time, the potential outcome of these developmental programmes remains questionable due to the serious psychological detachment of regional elites from the federal centre. In the view of the growing terrorist and separatist activity, the government plunges colossal resources to these areas just to keep the situation under control. In this context, its ability to achieve any genuinely positive integrative results remains to be seen in the near future.

The drive towards regional standardisation: ethno-federal problems

Another important problem lies in the fact that the federal reform, which initially aimed at the introduction of a greater transparency in the centre-regional and inter-regional relations, has ultimately evolved into a policy of regional standardisation (Turovsky 2005). The standardisation drive encompassed a large number of stringent initiatives that have had an adverse effect on the cultural diversity among the peoples of the Russian Federation. The ongoing processes of regional mergers and restrictive stance towards linguistic diversity have been of particular importance. In the first case, the centre attempted to enlarge a number of Russia's regions by liquidating pre-existing ethno-territorial formations and merging them with ethnic Russian regions. By December 2007, five such mergers had been accomplished, ultimately liquidating six ethno-territorial units. In the second case, Russia's authorities hoped to prevent the Federation's ethnic minorities from falling out of the all-Russian sphere of political influence and uniting around their distinctive linguistic and cultural lines.

As we can see, the drive towards regional standardisation has mainly encompassed Russia's ethno-territorial units that represented one main cause of centrifugal tendencies during the 1990s. In this context, it is important to bear in mind that all multi-national states, including Russia, contain, one way or another, the so-called nations without a state. Guibernau (1996, p. 101) distinguishes four types of nations without a state: nations whose cultural differences have been acknowledged by the host states, but have not been granted full institutional recognition (examples are Scotland and Wales before devolution); nations that have been granted a certain degree of autonomy and institutions by central governments (Catalonia and the Basque Country); nations that have been granted ethno-territorial status within a federation (Quebec); and, finally, nations that completely lack acknowledgement from the state which contains them (Kurdish people within different states amongst which the Kurdish land is partitioned).

In this light, one main problem associated with the regional standardisation drive is that, during Putin's presidency, Russia exhibited a consistent movement from the ethno-federal state category towards a league of states in which ethnic minorities, though recognised officially, lack any substantial institutional backing. Indeed, the Kremlin's policies towards national integration have slowly begun to evolve into a series of attacks on the political rights of Russia's nationalities. The State Duma consistently attempted to deprive the republican presidents of their presidential status by redesignating their posts as governors.[52] In 2005 the lower house denied Khanty-Mansi and Yamalo-Nenetsk Autonomous Okrugs the introduction of quotas on the indigenous nationalities for the composition of the regional legislatures.[53] All Muslim women have been forbidden from taking their passport photographs in *hijab*s – traditional Islamic headscarves (Turovsky 2005). However, two main dimensions of the national standardisation – the regional mergers campaign and restrictive linguistic policy – have had the most profound impact on the future of Russia's ethnic minorities and on the mode of inter-regional integration within the ethno-territorial dimension. The implementation of these policies exhibited the main traits of Russia's political style and was driven by the logic of economic and political expediency of interested elites. Moreover, the rigid administrative execution of such initiatives fomented serious legal battles and led to a simmering ethnic resentment.

Regional enlargement

The official arguments in favour of the regional merger process have been of an economic and legal-administrative nature. As of 2000, Russia had three administrative types of federation subject: 21 republics, six state-territorial formations termed *krais*, 49 *oblasts*, two federal cities; and 11 ethno-territorial formations – one Autonomous Oblast and ten Autonomous Okrugs (AO) (Kozlov 1996, p. 43). We have already argued that all these units, according to the 1993 constitution, represent equal parts of the Federation.

The principal legal-administrative difficulty related to this structure stemmed from the fact that nine out of the eleven AOs (with the exception of Chukotka AO) have also been administrative parts of other Oblasts or Krais. These larger formations have been commonly referred to as the 'mother' subjects, while the entire territorial construction has been termed a *matryoshka*, i.e. Russian doll which contains a number of other, smaller, dolls inside it.

Interestingly, populations of the *matryoshka* subjects, including their embedded AOs, had the right to elect legislative assemblies of the 'mother' regions. At the same time, the right to elect the Okrugs' authorities remained the prerogative of the AOs' residents. This created various political and administrative tensions. The leadership of Krais and Oblasts attempted to treat the AO authorities *on par* with its municipalities, while the latter, pointing at the constitutional equality with their 'mother' regions, strove to safeguard their political independence at the federal level (Kozlov 1996, p. 45).

Moreover, all AOs represented severely underpopulated territories. Indeed, while the neighbouring 'mother' subjects have been homes to over a million citizens, the AOs could hardly boast 100,000 inhabitants. In addition, the recent census demonstrated that in most these units, Russian ethnic majorities have long outnumbered the indigenous populations.

The final, and perhaps the most important, argument in favour of the AOs liquidation has been that these regions were economically backward and, therefore, had to unite with the territories of a stronger developmental and investment potential. Indeed, in most ethno-territorial units, the percentage of local revenues in the regional budgets, termed as *budgetary sufficiency*, was considerably lower than in their 'mother' subjects. Thus, many politicians insisted that, due to their unimpressive economic conditions and serious demographic problems, such territories do not have much electoral or political weight (Polin 2005).

Table 7.1 demonstrates the demographic situation in the five mergers already accomplished.

At the same time, despite these economic and legal-administrative incentives, the regional enlargement has not been a Rikerian 'coming together' grass-root process, in which certain territories voluntarily unite to achieve various economic and political benefits (Stepan 1999). Rather, this process has been implemented in Russia's arbitrary political style. Thus, the mergers have been imposed by the federal centre 'top down' often against the background of ethnic resistance, and mostly in accordance with the interests of ruling elites.

Interestingly, Russia's legislation theoretically grants local population with extensive rights during the merger process and makes sure that all the involved parties, including the populace, agree to the deed of merger. The law on 'The order of inclusion of a new subject into the Russian Federation and formation of a new subject of the Russian Federation within its structure', passed by the Federal Assembly in December 2001, assigns the enlargement initiative to the interested regions, which, following positive votes on referenda in their respective territories, collectively submit their proposals to the president. The president, in turn, passes these proposals to the Federation Council, State Duma, and the government – which then conducts a series of meetings and consultations to draft politico-institutional solutions to the proposed merger. When all the involved parties reach an agreed compromise, the president introduces the project of a new law on formation of the new federation subject, which outlines the subject's name, status, borders, and the structure of federal institutions within its borders. Finally, both houses of parliament discuss and ratify the law.[55] This system is somewhat similar to the one in Switzerland, where a chain of cantonal referenda, including separate municipal polls on the borders of a potential new canton and a national referendum on its recognition, precede any decision on the establishment of a new territorial unit (Fleiner 2002, pp. 107–8).

Despite the fact that the legal system in Russia was designed to preclude the centre from arbitrarily dividing the country's territories, the style of implementing the merger policies has remained traditional. Thus, the Kremlin's lead in the process has surfaced during Putin's first term in office and continued thereafter. The idea to form the new Perm Krai by merging Komi-Permyak AO and Perm Oblast initially belonged to the federal government and the process was conducted with an active deployment of administrative mechanisms (Goode 2004).

Table 7.1 Demographic and economic conditions within the enlarged territories[54]

Region	Merger date/new region name	Initial formation date	Total population	Budgetary sufficiency in 2004, %	Percentage of indigenous population
Krasnoyarsk Krai	1 January 2007/Krasnoyarsk Krai	7 December 1934	2,900,000	105	Russian ethnic region
Taimyr AO	1 January 2007/Krasnoyarsk Krai	10 December 1930	39,400	16.6	21.6
Evenk AO	1 January 2007/Krasnoyarsk Krai	10 December 1930	19,000	14.1	21
Chita Oblast	1 March 2008/Zabaikalskii Krai	26 September 1937	1,100,000	48.6	Russian ethnic region
Aginsk-Buryat AO	1 March 2008/Zabaikalskii Krai	26 September 1937	30,000	27.1	59
Irkutsk Oblast	11 October 2005/Pribaikalskii Krai	26 September 1937	2,500,000	82.1	Russian ethnic region
Ust-Orda Buryat AO	11 October 2005/Pribaikalskii Krai	26 September 1937	143,700	14.9	36
Kamchatka Oblast	1 July 2007/Kamchatskii Krai	10 January 1956	360,000	57	Russian ethnic region
Koryak AO	1 July 2007/Kamchatskii Krai	10 October 1930	23,800	28.9	26.7
Perm Oblast	1 December 2005/Permskii Krai	3 October 1938	2,800,000	43	Russian ethnic region
Komi-Permyak AO	1 December 2005/Permskii Krai	26 February 1925	136,000	23.1	59

Similarly, Putin openly suggested a territorial merger between Taimyr Autonomous Okrug, Krasnoyarsk Krai, and Evenkiya during his visit to Krasnoyarsk connected to the financial-industrial conflict among these regions.[56] A large number of federal politicians, such as Sergey Mironov, Vladimir Zhirinovskii, Alexandr Veshnyakov, presidential envoys Viktor Cherkesov and Leonid Drachevskii, have begun to advocate a general trend towards regional enlargement.[57]

During Putin's second term, the Kremlin has begun to overtly impose regional mergers 'from above' by administrative means. In Irkutsk Oblast, large industrial enterprises used various administrative measures to intimidate their workers into voting in favour of the merger during the April 2006 referendum.[58] The refusal of governors Alexei Barinov and Vladimir Loginov of Yamalo-Nenetsk and Koryak Autonomous Okrugs respectively to support the merger of their regions with neighbouring territories led, in the view of many experts, to their dismissal and the subsequent launch of criminal proceedings.[59]

More importantly, economic determinism represented the main driving force behind the merger process, which often disregarded national-cultural considerations. The head of the Federation Council Committee on territorial questions Alexander Kazakov, supported by a researcher from the Ministry of Economic Development Aleko Adamesku, openly declared that the *modus vivendi* of Russia's territorial reform should derive primarily from economic considerations, and if necessary, at the expense of the national-cultural characteristics of the existent regions.[60]

In this context, the federal centre has been virtually 'buying-out' the willingness of the regional leaders to conduct the merger campaigns by redistributing various grants and subsidies. Assessing this situation, the governor of Irkutsk region Alexander Tishanin has openly declared that, since the federal centre has embarked on the development of Siberian territories, the most efficient way to receive the ongoing subsidies is to become an enlarged region which would fall into the list of the Kremlin's priorities.[61]

Indeed, the unification of Krasnoyarsk Krai, Evenkiya, and Taimyr was followed by a special presidential decree No. 412, which provisioned substantial federal transfers (in the region of 23 billion roubles) aimed at the development of the enlarged territory.[62] While the federal support for Komi-Permyak Autonomous Okrug and Perm Oblast was less impressive than that allocated for the Krasnoyarsk case, the AO was awarded a single subsidy numerically similar to some three years of federal transfers.[63] In the immediate aftermath of declaring the merger, Kamchatka and Koryakiya regions forwarded their 'wish-list' to the president requesting to build five electricity stations, to construct a network of roads in strategic regional routes and to build a new airport.[64] The governor of Ust-Orda Buryat Autonomous Okrug Valerii Maleev demanded an additional grant of 8 billion roubles for the development of the republican social sphere and agriculture.[65]

Russia's large corporations have also been interested in modification of the existing territorial and administrative system. This was particularly true of

Putin's first presidential term, when, in an attempt to expand their political influence over neighbouring territories, representatives of big business had actively
lobbied in favour of the idea to enlarge a number of Russia's regions. In Siberia
and the Russian north, such developments were often associated with the implementation of new mining plans and the further construction of oil and gas pipe-
lines. For example, *Lukoil* had a vested interest in a potential merger between
Nenetsk Autonomous Okrug and Arkhangelsk region due to its long-term intention to obtain mining rights for the Val Gamburtseva oil well.[66] Similarly, a
union of Ust-Orda Buryat Okrug and Irkutsk Oblast was in the interest of *Transneft*, which had plans to build the important Angarsk-Nakhodka pipeline along
the north shore of Lake Baikal.[67] The merger of Krasnoyarsk regions with its
Okrugs was in the direct interests of *Rosneft* and *RAO UES*. In the wake of the
merger, Putin pledged to issue a special decree, supporting both companies'
plans towards the construction of a new Vankor-Dixon pipeline and the comple-
tion of the existing Boguchan Hydro-Electricity Station.[68] The prospective
merger (spring 2009) between Chita Oblast and Aginsk-Buryat Autonomous
Okrug was intended to boost the local non-ferrous metals mining industry and to
attract investors. The federal government pledged to invest some 40 billion
roubles (US$1.5 billion) in the regional mining capacities following the merger
and grant the *Norilsk Nikel* enterprise with exclusive developmental rights within
the new territory. It does not come as a surprise that *Nornikel* had some serious
vested interests in the ongoing regional merger. The company planned to invest
over 100 billion roubles (US$4 billion) in the exploration and mining of copper
in the five strategic locations: Bystrinskii, Lugokanskii, Kulumdinskii, Solnosh-
chenskii, and Bugdainskii mines (Verkhoturov 2007).

Finally, a large number of Russian regional politicians have exploited this
situation in order to secure various political dividends. Some leaders hoped to
enhance their nationwide popularity; others planned to improve their relationship
with the Kremlin. Many governors planned to head the enlarged territories, and
thereby, expand their existing power levers. This particularly concerned the governors of ethnic Russian regions in the European part of the country. Moscow
federal city, for example, has wanted to annex the Moscow region. The newly
created federation subject would become a highly developed, almost self-
sufficient, state, accumulating over a third of Russia's impressive economic
capacities and a substantial electoral potential. This could grant the Mayor of
Moscow Yurii Luzhkov, who has always been an advocate of tight governmental control over the economic affairs in his region, with unrivalled powers.
Indeed, since May 2006, Luzhkov has been peddling the merger idea, playing
the 'Kremlin support' card.[69] A similar situation occurred over the St Petersburg
Federal City and Leningrad Oblast, as well as the Yaroslavl and Kostroma, and
Bryansk and Orel regions.[70]

Alexandr Khloponin gained the most substantial political dividends by facilitating the merger process in his region. Khloponin introduced the idea of the
'second industrialisation of Siberia' and pledged to save this territory from political annexation by China – a threat feared by the vast majority of Russian politi-

cians.[71] Khloponin has also claimed to help the resolution of the national demographic problem by attracting some 100,000 migrants from the Former Soviet Union states to assist in the execution of grandiose industrial projects associated with the merger plans. An identical situation occurred in the Irkutsk region with governor Tishanin.[72]

It now becomes clear that central attempts towards regional integration carried off by the means of regional mergers and liquidating ethno-territorial structures have fundamentally rested on multi-faceted fears and politico-economic expediency of the ruling elites. The fears primarily concerned the dangers of Russia's breakup along the ethno-federal lines, as well as the threatening lack of territorial integration within the political and economic dimensions. However, from a comparative-theoretical point of view, such fears seem over-exaggerated and policy implementation methods come across as dangerous.

Existing academic literature suggests that ethno-federal states are more likely to collapse if they contain a single ethnic federal region that enjoys dramatic superiority in population, economic potential, and cultural influence. Duchacek (1986, pp. 283–5) refers to such regions as 'leading elements', while Hale (2004, p. 167) brands them as 'core ethnic regions'. Hale (2004) statistically demonstrates that '*all* ethno-federations that have collapsed possessed core ethnic regions, whereas no ethno-federation lacking a core ethnic region has collapsed'. This finding further supports the Horowitz (1985) argument that subdividing regions generally tends to facilitate ethno-federal stability.

Post-Soviet Russia, as opposed to its predecessor, the Soviet Union, does not possess such a *leading element*. This, perhaps, answers the question of why the country weathered the wave of extreme sovereignisation during the 1990s and preserved its territorial integrity against the odds of the growing regionalism, the war of federal-regional legislations, and the virtual dismemberment of its fiscal-economic space.

From this point of view, the current attacks on ethno-territorialism accompanied by the liquidation of the existing units seem an over-exaggerated reaction. Such a reaction could also lead to some potentially dangerous consequences. Economic determinism, which has been the *modus vivendi* behind the regional mergers, is particularly noteworthy. Statistical analysis, conducted with the Soviet and Russian examples, proves that the most economically advanced regions tend to have a higher propensity for secession. Emizet and Hesli (1995), in their pioneering work on the disintegration of the Soviet state, conclude that those republics which were most economically developed seceded first. Hale (2000) in his comparative follow-up analysis confirms that wealthier regions tend to be the most separatist in a given union state, since they have the most to lose should another group take control over the resources. Treisman (1997) further demonstrates that post-Soviet Russia's regions, which have acquired solid fiscal and economic potential, higher export capabilities, and valuable natural resources, have been the most outspoken in secessionist claims.

An additional range of reservations stems from the central suggestions that the liquidation of administrative units with a relatively small size of population is justified by their merger with larger and more populated regions. However, while the enlargement of these regions could, in fact, be justified on economic and legal-administrative grounds, the size of the population or territory does not play a large role in terms of the quest for national self-determination and nationalism, and could therefore not serve as a reason for liquidating these territories. Duchachek (1986, pp. 71–3) points out that the 1960s saw the emergence of a large number of tiny nations around the world. In 1967, for example, the 15-mile long Caribbean island of Anguilla (with just 5,000 inhabitants) decided to form an independent state instead of joining the projected federation of Saint Kitts-Nevis-Anguilla. Nauru Island proclaimed its independence in 1967 with just 3,100 inhabitants; Maldives obtained its independence in 1965, covering a total area of 112 square miles with the total population of 104,000 people.

These theoretical stipulations become ever more relevant in that the initial resistance of the indigenous population of almost all these regions to the alteration of their territorial borders undermined their commitment to the new structures. In 2001, ethnic Evenkiya decided against the idea of merging with Krasnoyarsk Krai and Taimyr Autonomous Okrug.[73] Political battles surrounding a potential merger between Yamalo-Nenetsk, Khanty-Mansi Autonomous Okrugs and Tyumen Oblast have also been well documented.[74] The Buryats' resistance to the merger with the Irkutsk region led the regional governor to liquidate a number of municipal units, in which the opposition seemed the most prominent.[75]

These examples demonstrate that, apart from economic considerations, attention must be paid to the ethical and ideological basis of the merger process. The question of who stands behind this drive – interested corporations, federal authorities, zealous regional leaders or real grassroots popular movements – should not be the last issue on the agenda. More importantly, economic determinism, which may well lead to the emergence of politically emboldened self-sufficient territorial units with little grassroots commitment to the central state, could initiate undesirable political tensions.

Restrictive language policy

Problems related to linguistic and cultural determination further complicated the process of regional integration in Russia. A range of restrictive language policy changes evoked ferocious legal battles in which the centre was occasionally forced to retreat. The seriousness of these disputes, as I will show later, has led to a never-ending debate on the limits of linguistic self-determination in Russia. In this light, Anthony Smith (1995, p. 51) observed that 'the fires of nationalism are never quenched', and neither tactical retreats nor coercive policies are likely to silence the perpetual bargaining between the centre and its ethno-territorial units. Indeed, despite various central pressures, Russia's ethnic minorities have never stopped voicing their claims to further linguistic independence.

Legal battles surrounding the wishes of the Tatar authorities to use the Tatar language in republican public offices on an equal footing with Russian underscored the futility of central attempts to suppress the manifestations of linguistic nationalism by administrative means. In particular, the centre demanded that Tatarstan changes Article 108 of its Yeltsin-era constitution, which required electoral candidates for the republican presidency to be Tatar-speaking citizens of the Republic. We have already argued that, from the point of view of the federal centre, this law practically denied a number of important freedoms to the rest of the country's population. The document ensured that even among the republic's citizens, only representatives of the Tatar nationality could fully participate in regional politics (Gorenburg 2001). It is also important that all Tatars (40 per cent of whom reside outside the republican borders) had the opportunity to participate in the political processes of other regions, while those of non-Tatar nationality (and there are over 50 per cent of such people residing in the republic) were virtually forbidden from taking official positions in Tatarstan.[76] A similar situation occurred in other republics: Buryatiya, Yakutiya, Chuvashiya.[77]

Even though Tatarstan made a serious concession to the federal centre in this sphere by introducing an amended version of its constitution in May 2002, the republican struggle for linguistic autonomy has not ended. Following the abolition of the linguistic requirement in Article 91 of the new constitution,[78] the already discussed 2007 bilateral Treaty reinstated the requirement for the republican presidential candidates to be fluent in both Russian and Tatar. In addition, the Treaty allowed Tatarstan's residents to carry specific nationality paper inserts in their passports written in the Tatar language and carrying the symbols of the republic.[79]

It is also important to mention that the new version of the Tatar constitution encapsulated a number of important concessions aimed at preserving the Tatar language. Article 8 of the new document states, for example, that both Tatar and Russian are official languages of Tatarstan and have equal usage in official institutions of the republic. Similarly, all official documentation is to be released in both languages. Other constitutional provisions, such as those in Article 56, guarantee the right to study the Tatar language in educational institutions and the right to read degrees in either Tatar or Russian.[80] These experiences have been repeated in a large number of ethnic regions of Russia's North – Evenkiya, Yamal, and Yakutiya being prime examples.[81]

Further central attempts at restricting regional linguistic self-determination ignited a simmering hostile reaction within certain ethnic republics. Of particular importance have been the wishes of the Tatar authorities to transfer the national alphabet from Cyrillic to Latin. This initiative was first formalised by the 1999 Tatar law 'On the restoration of the Tatar Alphabet on the Basis of the Latin script'. The federal authorities understood that the issues of language and alphabet might remove Tatarstan from the all-Russian sphere of cultural influence and grant the republic moral and socio-political justifications for further independence. The experience of other countries supports these concerns. In the Indian state of Assam, both Bodos and Tiwas tribes pressed for further territorial and political autonomy by military means, beginning their struggle with the demands to transfer their languages from the traditional Assamese script to Roman. This partly led to the introduction of the emergency act in 1975 (Rath 1984, pp. 74–5; Kimura 2003, pp. 236–7).

The Tatarstan policies towards the alphabet changes evoked serious legal battles and alarming social consequences. The previous version of the federal Law on 'the Languages of the Peoples of the Russian Federation', which was adopted on 25 October 1991, has been amended with the enactment of Law No. 165-FZ from 11 December 2002.[82] This new document established Cyrillic alphabet as the only legal script used for written languages within the Russian Federation and made the adoption of any alternative alphabets the prerogative of the federal centre. The conflict was concluded with the final ruling of Russia's Constitutional Court, which, in response to the demand of Tatarstan Constitutional Court from 24 December 2003 to rule Law 165-FZ unconstitutional, provided a virtuoso defence for the amendments. Its judgement from 16 November 2004 has instituted the Cyrillic alphabet as the only means of expressing written languages within the Russian Federation.[83] In addition, Law No. 53-FZ 'on the State Language of the Russian Federation' enacted on 7 June 2005 made Russian the sole official language of the Federation.[84]

These legal acts had an immediate adverse effect on the cultural development of certain nations. While Tatarstan harboured a silent resentment to these rulings, the political elites in Kareliya remained overtly critical. The indigenous population of the republic has a distinct cultural heritage and a language which belongs to the Finno-Ugric family and is written in the Roman script. The current legal arrangement has not affected Karelian because the regional authorities have so far failed to grant the language an official Republican status. At the same time, the ruling of the Constitutional Court precludes the language from obtaining such a status, as this step would involve its transfer to the Cyrillic script (Klimentev 2004).

It is important to mention again that the Kremlin's fears of losing control over the Russian linguistic space – that largely fuelled the extant inter-regional standardisation drive – have been seriously exaggerated. Academic literature suggests that any language develops in accordance with the internal logic of social processes. 'Street language', or the most commonly deployed media of communication, often prevails against the wishes of political elites. In Turkey, for example, colloquial Turkish has gradually replaced Ottoman Turkish deployed by the country's upper classes (Aydingun and Aydingun 2004, pp. 417–19). Similarly, Afrikaans – the language naturally used within the South African immigrant communities – emerged as the main medium of public communication following decades of widespread condemnation by the Dutch and English-speaking elites (Gilliomee 2004). The nation-building process in France and Italy was closely bound up with the emergence of local vernacular languages rather than with the promotion of Latin; and the contemporary development of the French language follows the 'street', rather than the guidelines established by the *Académie Française* (Safran 1999; Safran 2004, p. 7).

From that point of view, any attempts to impose various linguistic initiatives 'top down' are likely to have limited effects. More importantly, the adoption of restrictive policies could contribute to the growth of local nationalism, resentment, and the eventual mobilisation of regional ethnic minorities around their native languages. Legislation enacted in Sri Lanka in 1956 to establish Sinhalese as the only official language generated massive demands by Tamils for autonomy (Safran 2004, p. 5). Similarly, the attempts at forceful Arabisation of Algeria provoked protests of the ethnic Berbers and contributed to the emergence of social unrest (Benrabah 2004, pp. 73–4). The restrictive linguistic policy of the Quebec government fomented a hostile reaction across Canada. It led first to the emergence of the English language 'liberating' movements, such as the Quebec Alliance or Canadians for Linguistic Justice, and second to the adoption of retaliatory measures by the governments of the English speaking provinces (Cardinal 2004, pp. 82–4).

Depoliticising linguistic issues seems the most pragmatic path towards sociopolitical stabilisation and ethnic harmony. Canadian Prime Minister Pierre Trudeau endorsed the Canada Official Language Act in 1985 that granted English and French an official status not because of any conception of justice but because he believed that people who used these languages have 'obtained the power to break the country' (Blattberg 2006, p. 607). For very similar reasons, 13 languages have been accepted as official in India, and English, despite being the tongue of the former colonial power, was promoted as a media of communication in both economic and political circles (Safran 2004, p. 3).

This brief comparative outlook has a direct bearing on the situation in Russia. First, the restrictions imposed on the use of other languages may accelerate an already simmering hostile reaction. Second, despite the promotion of local languages by ethno-regional elites, Russian still remains the main medium of communication within the regions, and from this point of view, has little to fear. This particularly concerns Russia's Northern and East Siberian territories – Evenkiya, Taimyr, Yamal, and Buryatiya.[85] According to the 2002 census, only 5.2 million people use Tatar, which is considered the second most popular Russia's language after Russian.[86] Similarly, nearly a quarter of all Chuvash residing within the Russian Federation considered Russian as their native language, and 15 per cent of the Chuvash residing within the Chuvash Republic could not speak their native tongue.[87] From that point of view, the rational choice of regional populations will invariably remain on the side of Russian as the language of career advancement, economic development, and practical communication. Treisman's (1997, p. 232) comparative study of regional separatism in Russia underpins this point. He concludes that there was no statistical evidence of any relationship between separatist activism and the proportion of titular nationality using its ethnic language or the proportion of school students studying in their native language.

In conclusion, we can clearly see that the central attempts at re-integrating Russia's regional space have faced a range of serious limitations. These primarily stemmed from the logic of the reforms, the means by which many policies have been carried out, and the range of complementary restrictive initiatives. Russia's ethno-territorial dilemmas and nationality issues have played a central role in the emergence of negative dynamics. In 1998, Graham Smith (1998, p. 1398) noted that 'far from being based on a recognition that the ethno-republics should be treated distinctively', central policy reflected the 'anarchy of the political market place' and showed 'little regard for developing a coherent nationality policy'.

Our discussion has demonstrated that little has changed during the Putin years. The Kremlin has actively deployed the 'political market' logic for pursuing its ethno-regional policies, which often took place at the expense of national sentiments of the Federation's nationalities. The results and implications of these policies are highly questionable. The viability of the new territorial structures is fundamentally backed by political and economic considerations of participant elites, while the grassroots commitments to these units remain tenuous. Similarly, no restrictions, or even retreats, by the federal centre were capable of silencing the struggle of certain nationalities for their linguistic, and therefore, cultural self-determination. Legal battles surrounding these issues proved serious and unremitting. It is also important that economic favouritism has seriously marred the pattern of Russia's industrial growth and undermined her chances for regional economic integration. These developments lead one to suggest that, unless the government begins considering the lessons of comparative experience and taking steps towards more transparent regional policies, positive results from the integrative initiatives will be a long time coming.

8 Conclusion

This work has discussed the origins and results of the federal reforms initiated by Putin in May 2000. It has investigated the background to the reforms, the newly established centre-regional balance of power, changes in the functioning principles of central institutions responsible for the formulation of federal policy, and alterations in the functioning principles of regional institutions tasked with the development of their relations with the centre.

Following analysis of federal developments during 2000–8, this account suggests that, despite the fact that the regions' influence in national politics declined, the reforms have not led to a complete upheaval in the regional system. This relates to the existence of differing cultural and societal constraints that impeded the implementation of structural reforms in their original form. In many cases, such constraints stemmed from the ethnic and territorial considerations that pressed for a greater structural relaxation and regional autonomy. In other cases, these conditions were determined by the nature of Russia's arbitrary and informal *political style* that diverted modernising socio-economic initiatives. In most cases, governmental policies have led to a range of unintended consequences, and both regional and central political actors operated the newly emerged structures in a fashion that subverted the initial reformist drive.

Having examined the origins of Putin's federal policy, I conclude that these developments stemmed from a complex interplay between the structure and the process of Russia's federalism established under Yeltsin. The structural relaxation of the Yeltsin era had begun to threaten the territorial integrity of the Russian state and set in motion social and political forces, which pressed for a greater territorial and functional integration at the federal level. Moreover, there were a number of factors, which related to some fundamental traditions of Russian history.

One indication of the historic continuity is the existence of the centralisation-devolution cycles dating back to the days of the emergence of the Russian state. From this point of view, Putin's reform aimed at the re-centralisation of federal relations represented a logical response to the overall devolution of the Yeltsin era. With regard to the Yeltsin politics, the objective weakness of the regional institutional system established during that period led a number of politicians to search for ways to rectify these shortcomings as early as 1996–7. Various forces in Yeltsin's administration were elaborating on ways of reinforcing the power of the presidential representatives, possible means of removing the governors from office, the tactics of reforming the Federation Council, and the methods of reinstating state control over the fiscal and inter-budgetary policies. The ideas which emerged were very close to Putin's plan of reform. However, all attempts made during the Yeltsin era to restructure the federal system by measures implemented on a 'top down' basis were unsuccessful due to the lack of political consolidation at the centre and its embattled authority.

More importantly, there were reciprocal developments on the part of the regions, which sought to redesign the existing federal arrangement towards a greater symmetry and stability. The OVR movement headed by Luzhkov, Primakov, Yakovlev, and Shaimiev advocated reform of federal relations on a very similar basis to that proposed by the Kremlin. However, the capacity of the regions to form a single coalition advocating their interests at the federal level proved to be very weak, which made the attempts of the OVR to reform the Federation similarly unsuccessful. Scarcely less important was the fact that the centralisation of political power was widely supported by the electorate both in the regions and at the centre. These factors lead to the suggestion that Putin's proposed reforms had a solid institutional and socio-political basis that had been established during the previous regime.

If Putin's blueprints emerged from a general socio-political consensus, these reforms should have met with widespread acceptance. Indeed, the new policies halted the extreme regionalism, characteristic of the Yeltsin era, and linked the federation centre and the subjects legally and politically. In the wake of these developments, Russian federalism became less peripheralised, more symmetrical, and more centripetal. Most importantly, the federal constitution gained the status of a supreme and binding document for all the federation subjects, while the federal centre reclaimed the right to control its institutional branches in the regions.

At the same time, the new system of federal relations had a wide range of fundamental shortcomings such as the volatility of the new balance of power between the regions and the centre, the low level of institutional integration, and as a consequence, the low level of political stability. These problems mainly stem from the growing rift between the structure and the process of Russian federalism – a rift that has not diminshed during the course of Putin's regional reforms.

Thus, despite a range of positive developments, the reforms mainly resulted in a redistribution of powers supported by new rhetoric and not in a fundamental rethinking of federal relations towards adopting the essential functional principles of genuine federalism. One indication of this is the emergence of a new system of compromises that allowed the governors to retain most of the 'administrative resource' in their home territories and perpetuate a number of the old privileges for selected regions, which were masked under the guise of new verbal formulations. At the same time, any set of concessions given to the regions, normally in exchange for public support of the reforms, was counterbalanced by a set of limitations imposed on both the mode of implementation of these concessions and on the composition of actors involved. Moreover, in conceding the provision of certain privileges for certain territories the centre placed itself at the top of the new 'redistribution tree' and enhanced its control over the socio-economic activity of the regions – thus turning them into political 'clients'.

These developments ensured that the centre gradually obtained greater influence over regional political life and by the end of Putin's presidency became a dominant player in the centre-regional political game. In particular, the significant increase in the role of the centre has surfaced in the areas of inter-budgetary relations, the functioning of inter-regional and centre-regional institutions, the development of national parties and their regional activities, and the centre-regional migration of cadres. The regions, on the other hand, ceased to represent an independent pole of influence and became an integral part of the 'vertical' of state authority pursued by Vladimir Putin during his presidency.

It is important to mention that the initial ideas behind reforming the structure of federal relations invariably had democratising rhetoric based primarily on Western blueprints. Nonetheless, their implementation was dependent on the power struggle between the liberal and the conservative elite clans at the federal centre. The victory of the latter, whose representatives suggested that excessive privileges given to the regions would result in the emergence of uncontrollable poles of influence across the country, shaped the methods of implementing these reforms as centre-focused.

It is also significant that these constrictive methods were fundamentally based on Russia's specific *political style* that has represented one significant theme of the country's political culture. Thus, the resurfacing of such policy implementation methods was also consistent with theorising of many authors on that political culture could have a serious impeding influence on the democratisation process. Drawing on the anthropological writings of Anthony Cohen (1985) and Abner Cohen (1974), Welch (1993, p. 103) proposes viewing political culture as a certain tool-kit that is 'in some sense available for use, and may be voluntarily selected from on the basis of the requirements of new situations'. Swidler (1998) and Alexander (2000) develop this idea further, with Alexander distinguishing between various circumstances under which the reversal to the old cultural framework – as seen in the retrieval of the 'old tools' or, in Alexander's words, the *political culture thermidor* – is most likely. To his mind, if during a period of societal transformation continuous 'culture shocks' repeatedly bring unsuccessful results, the 'forces for the recreation of traditional culture will become apparent' (pp. 42–3). More importantly, if these traditional traits of political behaviour have been, in Brown's (2005, p. 187) terminology, *indigenously* established, they are likely to survive for some considerable period of time and surface within the conditions discussed above. These behavioural features may contribute to the survival of a certain *political style* that could travel from one institutional regime to another and outlive dramatic changes in belief systems. We have seen such developments within the conditions of Russia's troubled transformation. In this case, regional as well as central politicians understood and implemented many modernising initiatives in a manner consistent with their cultural beliefs and stylistic traits, which led to the acceleration of arbitrariness and informality of political conduct.

As a result of these institutional and societal dynamics, Russia has not made any significant steps towards the establishment of greater federal integration based on voluntary functional co-operation between regional and central institutional structures. Neither has it established a system of incentives which would encourage central and regional political societies to maintain their dialogue regardless of external pressures and changes in ruling political circles. Consequently, Russia's federal system moved to a position between 'guided' democracy and soft authoritarianism in terms of its political regime and exhibited characteristics of both a federal and a unitary state in terms of its institutional structure. The federal process, on the other hand, has obtained a distinctly decentralist vector, albeit has become adaptive and non-transparent in form. It is within this context that we should question the stability of the resultant institutional construction, the effectiveness of its functional parameters, the transparency of its decision-making processes and its general compliance with federal standards found in consolidated democracies.

Stability

Fundamental threats to the political stability of the new federal system are determined by four different factors: (1) the absence of genuine incentives for the governors to maintain the newly created federal order; (2) gubernatorial fear of potential coercion emanating from the federal centre; (3) retention of a significant 'administrative resource' by the regional leaders in their home territories; (4) dependence of the new federal system on the personal rating of the president and the composition of political elites at the federal centre.

The federal reforms failed to introduce a system of long-term genuine incentives which could motivate the heads of regional executives to sustain the newly established framework of federal relations. While such incentives should spring from well entrenched institutional co-operation between central and regional political elites, the Russian arrangement became deeply reliant on a multitude of short-term political concessions and economic subsidies granted to the regions by the federal centre. These include the right of a third electoral term for the most prominent heavyweight governors introduced at the beginning of the reforms, the centre's silent agreement to perpetuate a number of serious constitutional violations which existed in some 'strong' regions, gigantic investment programmes that the centre channelled into the most 'independent' territories in exchange for their compliance with national fiscal discipline, and various subsidies which originated from the general increase in social and fiscal demands placed on the regional budgets. Therefore, the potential of the centre to maintain the provision of these subsidies became the cornerstone of the stability of the new federal arrangement.

Moreover, as the subsidies grew, the centre began to secure firm control over their redistribution and ultimately increased its administrative potential in the regional sphere. The gradual implementation of a restrictive policy vector in the regions led to the situation in which the regional governors began to fear the possibility of coercion that could emanate from the federal centre headed by an energetic president. The official introduction of the legal possibility to evict the incumbent governors from office exacerbated these fears. From this perspective, gubernatorial loyalty to the existing institutional system seems rather volatile, resembling conformism rather than a voluntary consent to participate in the new framework of centre-regional dialogue.

Another perilous feature of the new institutional construction is manifest in the fact that despite the partial loss of authority and the status of national politicians, regional leaders retained a significant 'administrative resource' in their home territories. The most important indication of this is the central reliance on regional support during the parliamentary and presidential elections, during which the governors demonstrated their ability to control the outcome of voting. Regardless of the fact that the vector of this power is convergent with the policies of the federal centre at the time when central authority is strong, it could swing against the centre with a change of political circumstances in the Kremlin.

This suggestion seems even more valid given that the established balance of power between the regions and the centre is heavily dependent on the composition of political actors at the centre and in particular on the personal rating of the president. Above all, economic successes achieved during his presidency gave the public the illusion that the financial and institutional policies chosen by Vladimir Putin are the most suitable and that he should have 'carte blanche' in implementing them. However, the structure of the Russian budget demonstrates that the country's economic stability is 40 per cent reliant on high energy prices determined externally on the world markets. Significant changes in this volatile segment of the global economy could therefore seriously destabilise the institutional system by limiting the ability of the federal centre to honour the established agreements on regional subsidies. In this case, the governors could press for a significant alteration of the main institutional parameters of the existing federal arrangement and have sufficient authority to implement these demands. The same could be said with regard to Russia's rapid urbanisation and the spread of high techonology and information services into the regions. In this case, the growing socio-political and economic expectations of the general public would lead to further surfacing of societal pressures towards a greater regional autonomy and self-determination. This could represent another potential source of danger to the current centre-regional power construction.

Effectiveness

To some extent, Putin's federal reforms increased the effectiveness of centre-regional political dialogue. Within the conditions of extreme federal asymmetry created by the end of the Yeltsin period, the harmonisation between regional and federal legislation represented an important step forward in the centre-regional political relationship. Moreover, the antagonism that had characterised centre-regional relations during the Yeltsin era had given way to a more co-operative, even centripetal, style of communication.

At the same time, the effectiveness of federal relations depends on the ability of institutions responsible for the establishment of politico-economic dialogue between the centre and the regions and between the regions themselves to function constructively and independently. The range of these institutions includes the Federation Council, the State Council, its legislative counterpart, the system of national parties, and various organisations tasked with the development of certain areas of economic and social co-operation. Active participation of regional elites in these organisations can provide them with the incentives to create and sustain a stable federal framework and to search for the most optimal vectors of development of centre-regional political and economic dialogue.

This account has argued that Putin made some significant but insufficient attempts towards further integration of federal relations on the institutional level. The president established institutions of functional co-operation only with those segments of political society whose open resistance would doom the ideas of federal reforms to failure. These elites included regional governors, and representatives of regional legislatures and large business holdings responsible for generating the most substantial share of the country's GDP. At the same time, a wide range of organisations, the participation of which is essential for the development of an effective federal state, were left beyond the existing co-operative pact. The list of 'the excluded' was composed of the NGOs, the media, national political parties, and representatives of small and medium business. Scarcely less important remains the fact that co-operation with the 'chosen actors' had a strictly limited legal basis, thus failing to elevate these developments to a new, transparent level. One indication of this is the retention of dialogue within the regime of consultations leading to the conclusion of agreements mainly sustained by customs and not formally written rules. Moreover, the range of questions discussed concerns only those issues which are not of fundamental importance to the centre. Such an arrangement meant that any radical alternatives to Putin's ideas were blocked at the stage of preliminary hearings.

The system of national political parties represents perhaps the most important institution of centre-regional co-operation and the cornerstone of effective federal relations. However, the establishment of a strong national party system faced a number of constraints such as the partial regionalisation of the party space and attempts by the federal centre and representatives of big business to institute national parties 'from above'. While the regionalisation of party space mainly represented a hallmark of the Yeltsin era, the over-centralisation of this area of politics had grown into a serious problem by the end of Putin's first term.

The gap between the existing national parties and the population became evident with the 'privatisation' of these organisations by the federal centre and representatives of big business. For example, the aggressive promotion of the 'party of power' in the media and its overwhelming administrative potential transformed this organisation into a political force advocating the interests of the federal and regional bureaucratic elites. The role of the governors in this process was not insignificant: almost all regional leaders headed *Edinaya Rossiya* parliamentary lists in their respective territories. The emergence of *Rodina*, the new left-patriotic union headed by Glazev and Rogozin, was specifically designed by the Kremlin to split the Communist electorate during the December 2003 parliamentary elections. The subsequent formation of *Spravedlivaya Rossiya* underscored the Kremlin's determination to control Russia's party space by bifurcating the entire structure and creating two large controllable organisations. Russia's liberal parties could not take pride in having emerged as grassroots movements. SPS as a party created by important segments of the financial elite failed to appeal to the public and, before entirely losing its political influence by the December 2007 campaign, was widely regarded as a 'pocket' institution of big business. Similarly, *Yabloko*, despite having a much stronger political tradition and a better defined electoral niche, was misunderstood by the general public. A range of other small parties lost any chance of obtaining seats in the State Duma, which was increasingly dominated by Russia's four main organisations: *Edinaya Rossiya*, *Spravedlivaya Rossiya* (*Rodina*, Party of Pensioners, Party of Life), LDPR, and CPRF. Neither can these parties act as a mechanism for genuine societal consolidation. The subsequent resurgence of the national parties' activity in Russia's regions, which took place in the wake of enacting the new regional electoral model, has been eventually countered by the Kremlin's officials. Moreover, many such developments resulted in a direct participation of big business executives in these regional organisations with the aim of obtaining tangible economic benefits. Within these conditions, the existing political parties cannot act as effective institutions sustaining a genuine process of centre-regional political integration.

Beside the range of institutions of centre-regional politico-economic co-operation, independent inter-regional associations represent an important mechanism for improving the quality of centre-regional dialogue. By enabling the regional leaders to freely voice their opinion on the most important aspects of development of their territories, these organisations are indispensable to any effective federal relationship. The reforms led to increased inter-regional co-operation, though this process took place not through the establishment of independent associations of regional governors but through the traditional institutionalisation of the centre's predominance in this sphere. Inter-regional co-operation accelerated under the guidance of the federal centre: *polpredy* took over the leadership of previously created independent gubernatorial associations and even established a number of alternative organisations of inter-regional, international, and inter-district co-operation. At the same time, the new groupings soon made the role of the former independent inter-regional associations less prominent.

The loss of influence by independent gubernatorial movements demonstrated the fact that the governors have ceased to represent a self-sufficient pole of influence. They have been incorporated into the presidential 'vertical' of state power and made to promote decisions, ideas, and policies sent to the regions 'from above'. These developments have brought into question the ability of these institutions to reach effective decisions dictated by the real economic and political needs of the regions. The process of the regional mergers conducted on the basis of politico-economic expediency, further placed the process of inter-regional dialogue and co-operation under the central control.

Clearly, the functioning of the institutions of centre-regional and inter-regional co-operation was mainly sustained by the leading role of the federal centre, which was viewed by the Kremlin as a prerequisite for the successful development of the federal political process. However, if an independent functioning of such institutions does not represent the real substance of a federal system and the national parties can only be regarded as bogus formations sustained by the administrative resource of the centre and financial inflows of the oligarchs, such an institutional arrangement cannot ensure effective policy-making in the long term.

Transparency

Besides the inherent instability of the new federal arrangement and fundamental problems relating to its effectiveness, the political transparency of the reformed system generates a number of concerns. These relate mainly to central policy-making vis-à-vis the regions, the decision-making processes taking place within regional administrations, and the system of migration of regional cadres to national politics.

This work has suggested that, despite some alterations in central decision-making, the system has not become more transparent. During both Yeltsin's and Putin's presidencies the highest political leadership reached a number of private decisions without exposing them to public debate and scrutiny. In both cases, these processes were sustained mainly by custom and not by formally written rules. Indeed, the new system of interaction between central institutions responsible for the establishment of federal policy towards the regions was operating on the principle of an 'institutional power interlock', which made all these organisations directly dependent on the president, thereby granting him substantial arbitrary powers. Ultimately, this system reduced the scope for open inter-institutional dialogue, created the illusion of political stability, and as a result reduced the transparency of political processes at the highest level.

At the regional level, the increase of political transparency in the legislative process was counterbalanced by further convergence between big business and regional political power. On the one hand, it was thanks to the establishment of a number of institutions which were due to test the emerging legislation on the subject of compliance with federal standards, that the regional legislative process became more transparent and balanced. More importantly, the regional legislatures experienced a significant boost to their authority by gaining the right to approve regional budgets and control their implementation, by increasing the number of deputies employed in the legislatures on a full-time basis, and by electing 50 per cent of their representatives from party lists. It is also important that the new order instituted compulsory public discussions for any potential bilateral treaty signed between the regions and the centre.

On the other hand, the impact of migration of national big business into the leading administrative positions in the regions appears to be significantly less straightforward. This process took place as a consequence of the 'de-oligarchisation' campaign conducted by Putin at the centre and to some extent had a positive effect on the level of transparency of centre-regional dialogue. In contrast to the governors, whose main objective was to hold absolute authority within a clearly defined territory, big business had much wider national interests which were unthinkable without the establishment of a system of common rules for all the regions in their relations with the centre. Coincidentally, in this sphere the interests of the federal centre and big business converged, which facilitated the process of harmonisation of federal and regional legislation and ultimately increased the transparency of centre-regional political dialogue.

At the same time, the migration of large enterprise influence from the federal centre to the subjects brought the regions to the brink of a situation in which political power and business merge into a single indivisible complex. This process made the regional leaders reliant not only on central political trends but also on the disposition of forces within regional and national business elites. As a consequence, the transparency of regional administrative systems was marred by the 'privatisation' of political authority in the regions, business wars for the redistribution of political and economic resources, and the development of monopolistic trends undermining fair economic competition in the sphere of small and medium business. It is also important that the range of corrective policy initiatives adopted by the federal centre has led to the new, and at times more sophisticated, forms of relational informality between the two parties.

Furthermore, the introduction of the gubernatorial appointment system ultimately led to the consolidation of political elites within the regions. This was accompanied by the growing lack of transparency in the regional cadre policy and decision-making. Most governors were chosen behind the Kremlin's closed doors and voted by the regional legislatures without major discussions on potential alternatives. The resulting stagnation of cadres in the regional high power echelons has become striking. Many leaders have occupied their positions for over 15–18 years and their cabinets have some similar levels of longevity.

An area not without its problems is the process of migration of regional cadres to important positions in national politics. A basic level of political transparency in this sphere represents one of the most fundamental requirements for the effective functioning of a federal state. The fulfilment of such requirements guarantees a stable alteration of power by generating new national leaders and provides regional politicians with genuine incentives to participate in national politics.

At the same time, the institutional developments conducted in the framework of federal reform failed to follow this logic. One indication of this is the influx of the St. Petersburg political elite to the federal centre, which took place on a selective arbitrary basis rooted principally in loyalty and familiarity with the existing president and not in the objective merits of political talent and independent thinking. Moreover, the introduction of the gubernatorial appointment system, as well as the right given to a number of regional executive heads earlier to stand for a third term, deprived these governors of the incentive to pursue their careers at the federal level and impeded the emergence of young politicians in the regions by solidifying the authority of the existing elites. Similarly, the new method of formation of the Federation Council resulted in the promotion of pro-gubernatorial members of regional elites, well-established business executives, and Moscow officials, thus blocking the path for potential independent candidates. These developments disregarded the fact that the upper chamber should generate self-sufficient national politicians from the regional reserves.

Final suggestions

The system of federal relations created under Putin's leadership represented a step forward from the regionalism of the Yeltsin era. At the same time, all of the most serious shortcomings of the new arrangement stem from the fact that the Kremlin has placed the focus of the reforms on centralisation and the strengthening of state control. This has resulted in a substantial increase in the role of the centre in various spheres of centre-regional relations: inter-budgetary affairs, the functioning of inter-regional and centre-regional institutions, the development of national parties and their regional activities, and the centre-regional migration of cadres. At the same time, the ultimate re-centralisation of the federal model has not resulted in a fundamental tightening of real institutional processes. Regional elites preserved a substantial degree of autonomy within their respective jurisdictions and received a strong impetus for consolidation and strengthening. More importantly, the new institutional system, despite some of its modernising impulses, has not led to a fundamental restructuring of political actors' thinking on how federal institutions should be operated. Thus, the regional elites have begun to exercise the existing degree of informal freedoms with the view of exerting devolutionist moves in a disguised adapted fashion. This often led to the emergence of arbitrary shadowed politics and reaching various decentralising decisions on the ad-hoc non-transparent basis.

Bearing these developments in mind, the new institutional system experiences a number of problems in the spheres of stability, effectiveness, and transparency. The ultimate reduction of alternative poles of political influence restricts the ability of the supreme executive to balance between differing developmental approaches and vary its strategy when faced with unsuccessful outcomes. Moreover, such an arrangement places the entire burden of socio-political responsibility and decision-making powers on a small group of people at the centre and transfers the political process from the public sphere to informal bargaining. Similarly, the predominance of the centre brings into question the ability of those institutions responsible for the establishment of politico-economic dialogue between the regions and the centre to reach constructive and independent decisions. Finally, the dependence of regional leaders on central concessions and subsidies substitutes their open dialogue with the federation centre for political passivity and conformism.

In this context, it may be argued that the Russian Federation can only become a genuine federal democracy if the ultimate predominance of the centre in the functioning of its regional system gives way to a 'bottom up' approach to the process of federal integration. This means that the regional leaders must obtain genuine long-term political motivation for the maintenance of the existing federal system, and this motivation should not depend on current political trends at the centre. In order to transfer these requirements from theoretical grounds into practice it is necessary to revise the existing political style of implementing central policy ends. Practically, it is essential to ensure the provision of independent functioning for the institutions of centre-regional and inter-regional co-operation, the removal of fear of administrative coercion applied to the heads of the regional executives, the liberalisation of the activities of the national parties, the promotion of independent regional and federal media, and the separation of business and power on both the regional and federal levels.

However, in the current political climate, reform of overbearing state control can only be accomplished with the agreement of a very narrow circle of the federal elite to take steps towards a self-imposed power restriction and the liberalisation of various aspects of political life. We can only express the hope that the pursuit of such a liberalisation – perhaps partial and balancing within the limits of guided democracy – will offer a viable solution to the difficulties facing contemporary Russian society. This, in turn, may provide answers to the problems of stability, effectiveness, and the transparency of the existing federal system.

Notes

1 Methodology, theoretical considerations, and structure of the study

1 Putin's Address to the Federal Assembly, 8 July 2000, Offitsialnoe Internet Predstavitelstvo Prezidenta Rossii, available at http://president.kremlin.ru/text/appears/2000/07/28782.shtml; also see Vladimir Putin interview, *Nezavisimaya Gazeta*, 25 December 2000.
2 Stepan (1999, pp. 22–3) classified federations as 'coming together' and 'holding together' types.
3 'Realnaya Rossiya ne sovpadaet s edinoi: v nei pyat klassov i chetyre ideologii', *Vedomosti*, 6 May 2006.
4 'Pervyi raz v srednii klass', *Vremya Novostei*, 20 April 2007.
5 'Pravitelstvo mozhet otdat stabfond pensioneram', *Vedomosti*, 17 January 2007; 'Proshtrafilis!', *Nezavisimaya Gazeta*, 7 April 2006.

2 Russia's federalism under Yeltsin: the structure, process, and origins of Putin's reforms

1 The term constitutional-treaty, *konstitutsionno-dogovornaya*, entered the political lexicon quite early (see Rumyantsev 1995; Plyais 1998). Kahn (2000; see also Kahn 2001) was among those who highlighted the treaty-constitutional nature of the federation, thus emphasising the supremacy of treaties over the constitution.
2 However, this picture is schematic. For example, according to data from the Ministry of Trade and Economic Co-operation of the Republic of Tatarstan, in 1999 48 per cent of Tatars reside outside the Republican borders. Similarly, 70 per cent of the population in the republic of Adygeya are Slavic.
3 Mintimer Shaimiev interview, *Izvestiya*, 12 February 1999.
4 Putin's Address to the Federal Assembly, 8 July 2000, Ofitsialnoe Internet Predstavitelstvo Prezidenta Rossii. Online. Available at http://president.kremlin.ru/text/appears/2000/07/28782.shtml (accessed 20 April 2009).
5 'Ultimatum Ustinova', *Izvestiya*, 2 June 2000.
6 Ibid.
7 The Constitution of the Russian Federation, Chapter 1. The Fundamentals of the Constitutional System. Online. Available at www.gov.ru/main/page4.html (accessed 20 April 2009).
8 The Constitution of the Russian Federation, Chapter 1. 'The Fundamentals of the Constitutional System', Article 66.1 and 66.5; Chapter 3. 'The Russian Federation'.
9 That said, the Federation Treaty could only be regarded as an internal administrative document on the division of powers between the subjects of the Federation and the national centre, and not an international document establishing an independent federal state. All the signatories had already existed within the Russian Federation as subjects

of that Federation in accordance with the 1978 RSFSR Constitution. The impact of the discrepancies between the Treaty and the constitution was therefore not of a legal-institutional nature as such but of a moral and socio-historic substance, and was largely associated with the loss of the legitimacy of the pre-existing Soviet state and the general decentralising dynamics that surfaced within the centre-regional dimension, following the collapse of the USSR. More importantly, the new federal constitution superseded the Federation Treaty by stating that wherever the Treaty contradicts the constitution the latter takes precedence.

10 The Constitution of the Russian Federation, Chapters 1 and 3.

11 The Constitution of the Russian Federation, Part 3, Article 72.1(k).

12 President's Address to the Federal Assembly, 8 July 2000, Ofitsialnoe Internet Predstavitelstvo Prezidenta Rossii, http://president.kremlin.ru/text/appears/2000/07/28782.shtml (last accessed 20 April 2009).

13 The position of Vladimir Lysenko, the chairman of the State Duma committee on Federal Affairs, was close to such a solution. Lysenko (1997b, pp. 14–20; see also Lysenko 1997a and 1995) often called for the development of federal legislation that would institutionalise a degree of federal 'interference' in regional matters.

14 A number of regional politicians and academics, in particular Khakimov (1998 and 1997) and Toshchenko (1994) favoured such a solution. They suggested that Russia's regions represent historically and politically different entities and that it is not possible to draw up a single piece of legislation accommodating these differences.

15 Sverdlovsk Oblast was the first non-ethnic region to sign a bilateral treaty with the centre on 12 January 1996. The process was concluded on 16 June 1998 when Moscow Federal City obtained a similar document. By this time, 46 out of Russia's 89 regions had institutionalised their relations with the federal centre on the basis of bilateral treaties.

16 Petrov (2000b) observed these oscillations even within a shorter period of post-Soviet history.

17 The political competition between Alexandr Lebed and Valerii Zubov has been resolved with an active participation of central authorities. The Mayor of Moscow Luzhkov visited the region during the electoral campaign, and Moscow political commentator Sergey Dorenko supported Lebed in his analytical programmes. The centre has also turned a blind eye to the fact that Lebed's electoral fund exceeded by 33 times the legal electoral limit and comprised some US$2.3 million.

18 'Respublika Kalmykiya', *Kommersant Vlast*, 14 October 2002.

19 Data collected from the Official Website of the Russian Government, available at www.gov.ru and *Radio Liberty Radio Free Europe reports*.

20 'Kreml Obrugal Buryatov Poslednimi Slovami', *gazeta.ru*, 23 November 2001. Online. Available at www.gazeta.ru/2001/11/23/kremljobruga.shtml (accessed 20 April 2009).

21 Dmitrii Ayatskov interview, *Nezavisimaya Gazeta*, 15 September 1998.

22 Egor Stroev interview, *Rossiiskaya Gazeta*, 14 February 1998.

23 Yurii Luzhkov interview, *Parlamentskaya Gazeta*, 14 October 1998.

24 'Rossiiskie Regiony: Put v Novoe Tysyacheletie', *Rossiiskaya Federatsiya Segodnya*, No. 13, 1998.

25 'Primorskii Krainii', *Kommersant Vlast*, No. 6, 13 February 2001.

26 Interview with the President of Tatarstan M. Shaimiev. Online. Available at www.Shaimiev.ru/interviews/012 (accessed 20 April 2009).

27 *Obshchaya Gazeta*, 26 April 1999.

28 'Pervoe Polugodie 1999. Otnosheniya Tsentr-Regiony', *Vybory v Rossii*, May 2000. Online. Available at www.vybory.ru/sociology/polit_1half1999centreg.php3 (accessed 20 April 2009).

29 This support, however, was partially explained by Chubais' animosity with Berezovskii after the privatisation of *Svyazinvest* in 1997 and his plans to diminish the influence of the oligarchs in the Kremlin.

30 Interview with the Deputy Head of the presidential administration Oleg Sysuev, *Kommersant Vlast*, No. 17, 1999.
31 'Posredstvom Nakhrapa. Nazdratenko kak Simvol Vremeni', *Itogi*, No. 5, 2001.
32 'Vybory v Dumu: Suzhdeniya i Prognozy', *Nezavisimaya Gazeta*, 15 October 1999.
33 These views were often expressed in interveiws, television debates, and in the printed press. One such example is the article by M. Leontev (2004), in which he advocates the creation of a strong state with developed 'repressive apparatus' as a basis for the further modernisation of Russia. This article evoked heated public debates in the media and a number of observers such as Liliya Shevtsova (2004) and Vladimir Ryzhkov (2004) responded to it in *Izvestiya*. For further extensive reading on this subject see Lukin (2001).
34 Working Documents of Mytishchi *Raion* Administration, Moscow Region (unpublished).
35 VTsIOM, 31 May 2000. Online. Available at www.wciom.ru/?pt=42&article=482 (accessed 20 April 2009).
36 Eduard Rossel interview, 'Rossiyu Neobkhodimo Razdelit na Respubliki', *Izvestiya*, 19 June 2001.
37 Alexander Lebed interview, *Nezavisimaya Gazeta*, 2 June 1998.
38 Farid Saphiullin interview, *Radio Liberty Russia*, 1 June 2000.
39 Members of Yeltsin's financial-political clan were often addressed as the 'family' due to the fact that Russia's first president consulted his daughter Tatyana Dyachenko on most important political issues. She subsequently married the head of the presidential administration Valentin Yumashev.
40 Boris Berezovskii, 'Open Letter to the President', *Kommersant Daily*, 30 May 2000.
41 One of the most prominent scandals concerned the diversion and subsequent laundering of $10 billion of International Monetary Fund loans. The officials who fell under the investigation included: Tatyana Dyachenko, Yeltsin's daughter and closest adviser, Anatolii Chubais, Yeltsin's former chief of staff and finance minister, Oleg Soskovets, the former deputy prime minister of Russia, Alexandr Livshits, a former finance minister, and Vladimir Potanin, former deputy prime minister of the Russian Federation.
42 S. Ivanov, the Secretary of the Security Council, interview, *Komsomolskaya Pravda*, 8 June 2000.
43 *Radio Free Europe Radio Liberty Report*, 23 June 2000.
44 Press Conference by President W. Clinton and President V. Putin, St. Georges Hall, the Kremlin, Moscow, Ofitsialnoe Internet Predstavitelstvo Prezidenta Rossii. Online. Available at http://president.kremlin.ru/text/appears/2000/06/28760.shtml (accessed 20 April 2009).
45 *BBC World Service*, 21 July 2000.
46 Otnoshenie k Prezidentu posle Istorii s Arestom Gusinskogo, Regionalnyi Sotsiologicheskii Monitoring Fonda Indem. Online. Available at www.indem.ru/idd2000/arpi/07_12_2000/otchet14.html (accessed 20 April 2009).
47 'Lider Yabloka Prizyvaet Vesti Zhestkii Dialog s Vlastyu', *Segodnya*, 16 June 2000.
48 'Demokraty Reshili Eshche Nemnozhko Obedinitsya', *Russkaya Mysl*, 29 June 2000.
49 'Gospodin Kotenkov kak Zerkalo Diktatury Zakona', *Nezavisimaya Gazeta*, 6 June 2000.
50 'Boris Nemtsov: Senatoram Luchshe ne Upryamitsya', *Izvestiya*, 03 June 2000.
51 *Expert*, No. 5, 3 July, 2000; Doklad Levogo Informatsionnogo Tsentra, *Analiticheskaya Gruppa LITS* 9 July 2000. Online. Available at www.levy.ru/news/analytics (accessed 20 April 2009).
52 'Otklonit bez Soglasitelnoi Komissii', *Svobodnyi Kurs, Barnaul* No. 27 (472), 6 July 2000.
53 *Interfax*, 31 May 2000.
54 Putin interview, *Nezavisimaya Gazeta*, 26 December 2000.
55 Ibid.

3 The centre and the regions: a new balance of power

1 'Edinyi Nalog Nravitsya Vsem. Osobenno Sverkhbogatym', *Parlamentskaya Gazeta*, 12 April 2001.
2 'Putin: Big Tax Cut for Small Business', *The Moscow Times*, 29 March 2002.
3 'Sovet Federatsii Odobril Byudzhet-2004', *Nezavisimaya Gazeta*, 10 December 2003.
4 'Byudzhet 2004: Regionam Obeshchayut Pomoch', Tsentr Izucheniya Regionalnykh Problem, 13 October 2003. Online. Available at www.reg-center.ru/finance/0310/1. shtml (accessed 20 April 2009).
5 Ibid.
6 Ibid.
7 Author's interview with N.M. Rebchenko, Deputy Governor of Moscow Region.
8 VTsIOM polls constantly registered popular demand for consolidating the country's positions on the international arena. Most importantly, from the August 2002 poll Levada concluded that Russians viewed Putin's activities on the international arena as his most impressive achievement. See Yurii Levada, 'Sotsialno-politicheskaya situatsiya v Rossii v Avguste 2002 goda po dannym oprosov obshchestvennogo mneniya', www.wciom.ru/?pt=42&article=206, last accessed 20 April 2009.
9 'Ne Zhenikh', *Izvestiya*, 17 May 2002.
10 Putin's Television Interview with Citizens of the Russian Federation, 18 December 2003. Online. Available at http://president.kremlin.ru/text/appears/2003/12/57398. shtml (accessed 20 April 2009).
11 Central economic executives publicly admitted this situation at a meeting in the Federation Council. See 'My naobmanyvali ludei na 6.5 trillionov rublei', *Kommersant Daily*, 26 October 2002.
12 Ukaz Prezidenta RF ot 13.05.00 No. 849 'O polnomochnom predstavitele Prezidenta Rossiiskoi Federatsii v Federalnom Okruge', *Rossiiskaya Gazeta* 14 May 2000.
13 Federalnyi Zakon RF 'O Vnesenii izmenenii i dopolnenii v stati 7 i 9 Federalnogo Zakona "O Militsii"', Moscow, 04 August 2001. Online. Available at http://document.kremlin.ru/index.asp (accessed 20 April 2009).
14 Federalnyi Zakon RF 'O Militsii' ot 18 February 1993. Online. Available at www. uvd.lipetsk.ru/law/990301.htm, see also http://hro.org/docs/rlex/militia/index.htm (accessed 20 April 2009).
15 Federalnyi Zakon RF 'O Vnesenii izmenenii i dopolnenii v stati 7 i 9 Federalnogo Zakona "O Militsii"'.
16 Zakon 'O Vnesenii izmenenii i dopolnenii v Federalnyi Zakon ob "Obshchikh printsipakh organizatsii zakonodatelnykh i ispolnitelnykh organov gosudarstvennoi vlasti subektov Rossiiskoi Federatsii"', *Sobranie Zakonodatelstva Rossiiskoi Federatsii*, 31 July 2000, pp. 6075–80.
17 'Postanovlenie Konstitutsionnogo Suda RF po delu proverki konstitutsionnosti otdelnykh polozhenii Federalnogo Zakona "Ob obshchikh printsipakh organizatsii zakonodatelnykh i ispolnitelnykh organov vlasti subektov RF" v svyazi s zaprosom Gosudarstvennogo Sobraniya Respubliki Sakha i Soveta Respubliki Gosudarstvennogo Soveta Respubliki Adygeya, 4 April 2002', *Sobranie Zakonodatelstva Rossiiskoi Federatsii*, 15 April 2002, pp. 6553–80.
18 'Tri Putinskikh Udara', *Kommersant Daily*, 20 May 2000.
19 Federalnyi Zakon No. 159-FZ ot 11 December 2004 'O Vnesenii Izmenenii v Federalnyi Zakon "Ob Obshchikh Printsipakh Organizatsii" i v Federalnyi Zakon "Ob osnovnykh garantiyakh izbiratelnykh prav"', *Rossiiskaya Gazeta*, 15 December 2004; Ukaz Prezidenta Rossiiskoi Federatsii No. 1603 ot 27 December 2004 'O Poryadke rassmotreniya kandidatur na dolzhnost vysshego dolzhnostnogo litsa subekta Rossiiskoi Federatsii', *Rossiiskaya Gazeta*, 29 December 2004.
20 Zakon 'O poryadke formirovaniya Soveta Federatsii Federalnogo Sobraniya Rossiiskoi Federatsii', *Sobranie Zakonodatelstva Rossiiskoi Federatsii*, Vol. 20, 15 May 2000, pp. 4319–24.

21 'Ves Sovet Federatsii', *Kommersant Vlast*, No. 7, 26 February 2002.
22 'Takaya Ptitsa ne Letaet', *Nezavisimaya Gazeta*, 25 December 2003.
23 Federalnyi Zakon 'O vyborakh deputatov Gosudarstvennoi Dumy Federalnogo Sobraniya Rossiiskoi Federatsii ot 18 May 2005', *Rossiiskaya Gazeta*, 25 May 2005.
24 President's Address to the Federal Assembly, Moscow, 3 April 2001. Online. Available at http://president.kremlin.ru/text/appears/2001/04/28514.shtml (accessed 20 April 2009).
25 Federalnyi Zakon RF 'O vnesenii izmenenii i dopolnenii v statyu 4 Federalnogo Zakona Rossiiskoi Federatsii "Ob obshchikh printsipakh organizatsii zakonodatelnykh i ispolnitelnykh organov subektov Rossiiskoi Federatsii"', Moscow, 24 July 2002. *Rossiiskaya Gazeta*, 25 July 2002.
26 See Article 4.9 of Federalnyi Zakon RF 'Ob osnovnykh garantiyakh izbiratelnykh prav i prava na uchastie v referendume grazhdan Rossiiskoi Federatsii', Moscow, 25 July 2002, available at the official website of the Central Electoral Committee, www.cikrf.ru/_3/zakon/zakon67_02/Fz-67.htm (accessed 20 April 2009).
27 A number of politicians from the Regional Legislature of Stavropol Krai admitted in conversations that such an innovation made legislative processes more democratic.
28 President's Address to the Federal Assembly, 18 April 2002.
29 Federalnyi Zakon 'O printsipakh i poryadkakh razgranicheniya predmetov vedeniya i polnomochii mezhdu organami gosudarstvennoi vlasti Rossiiskoi Federatsii i organami gosudarstvennoi vlasti subektov Rossiiskoi Federatsii.' Moscow, 24 June 1999. Online. Available at http://document.kremlin.ru/index.asp (accessed 20 April 2009).
30 Ibid.
31 Ibid.
32 'Vopros Nomera: Schitaete li Vy Nyneshnie Mezhbyudzhetnye Otnosheniya Normalnymi?', *Rossiiskaya Federatsiya Segodnya*, No. 3, 2002.
33 A similar programme was adopted for Bashkortostan.
34 'Shaimiev – Putin', *Nezavisimaya Gazeta*.
35 'Byudzhet 2004: Regionam Obeshchayut Pomoch', Tsentr Izucheniya Regionalnykh Problem.
36 'Bednye Rodstvenniki', *Novye Izvestiya*, 15 March 2007.
37 'Tochki Rosta Rossiiskoi Ekonomiki', *Vedomosti*, 31 October 2007.
38 'Regiony Priblizyat k Dengam', *Rossiiskaya Gazeta*, 1 February 2008.
39 'Investfond Poteryal Dva Goda', *Rossiiskaya Gazeta*, 8 February 2008.
40 Federalnyi Zakon 122-FZ 'O vnesenii izmenenii v zakonodatelnye akty RF i priznanie utrativshimi silu nekotorykh zakonodatelnykh aktov RF v svyazi s prinyatiem federalnykh zakonov "O vnesenii izmenenii i dopolnenii v Federalnyi Zakon 'Ob obshchikh printsipakh organizatsii zakonodatelnykh i ispolnitelnykh organov gosudarstvennoi vlasti subektov RF' i 'Ob obshchikh printsipakh organizatsii mestnogo samoupravleniya v RF'"', *Sobranie Zakonodatelstva Rossiiskoi Federatsii*, Moscow, 22 August 2004, p. 2711.
41 Rukovoditeli Pyati Regionov Napomnili Putinu o ego Obeshchaniyakh, *polit.ru*, 28 July 2004. Online. Available at http://polit.ru/news/2004/07/28/put.popup.html (accessed 20 April 2009).
42 Ibid.
43 'Byudzhet 2004: Regionam obeshchayut'.
44 'Shvidkin do Deputatov ne Doshel', *Vremya Novostei*, 15 February 2001.
45 'Gubernator Rossel Vynes Preduprezhdenie Prezidentu', *Kommersant Daily*, 15 December 2000.
46 Author's interview with N.I. Rebchenko.
47 One positive development in this context is that the system of federal inspectors is quite sporadic and often unites two or three regions on a random basis. More importantly, it disregards the traditional ethno-territorial method of governance, thus making the administrative structure of federal inspectors' offices more equitable. For

example, Kabardino-Balkariya was 'merged' with Mineralnye Vody, Sakha Yaku-tiya with Magadan Oblast, Khakasiya with Tyva, Koraik and Chukotka Autonomous Okrugs with Kamchatka Oblast.

48 Ukaz Prezidenta Rossiiskoi Federatsii No. 773, 2 July 2005 'Voprosy Vzaimode-istviya i koordinatsii organov ispolnitelnoi vlasti subektov Rossiiskoi Federatsii i territorialnykh organov federalnykh organov ispolnitelnoi vlasti', *Rossiiskaya Gazeta*, 8 July 2005.

49 President's Address to the Federal Assembly, Moscow, 18 April 2002. Online. Available at http://president.kremlin.ru/events/510.html (accessed 30 September 2008).

50 'Gubernatory Otkazyvayutsya ot Privelegii', *Kommersant Daily*, 10 July 2001.

51 Constitution of the Republic of Tatarstan, Part 1, Article 1.

52 Constitution of the Republic of Bashkortostan, Preamble.

53 The Federation Council blocked the initial Treaty draft voted by the State Duma on 7 February 2007. Given the political subordination of the Council, this could not have occurred without Kremlin pressure.

54 'U Mintimera Shaimieva otbirayut podarok', *Nezavisimaya Gazeta*, 21 February 2007.

55 'Defile suverenitetov', *Vremya Novostei*, 6 March 2007.

56 'Vtoroe vzyatie Kazani', *Nezavisimaya Gazeta*, 19 December 2006.

57 'Defile'.

58 'Bez nalogov', *Vedomosti*, 6 June 2007.

59 'Dogovor na povyshennykh tonakh', *Kommersant Daily*, 15 November 2006.

60 President's Address to the Federal Assembly, Moscow, 3 April 2001. Online. Available at http://president.kremlin.ru/text/appears/2001/04/03/0000_type-633724type82634_28514.shtml (accessed 20 April 2009).

61 *Interfax*, 17 November 2004.

62 *RIA Novosti*, 19 May 2004.

63 'Smotr Administrativnogo Resursa', *Vremya MN*, 11 February 2003.

64 *RIA Novosti Ural*, 24 October 2002. Online. Available at http://ural.rian.ru/news.html?nws_id=33544 (accessed 20 April 2009).

65 'Tatarstan v Grazhdanskom Brake s Rossiei', *Nezavisimaya Gazeta*, 23 April 2002.

66 Article 1.1 of the document states that 'sovereignty is the integral qualitative charac-teristic of the republic and is expressed through the execution of full power in the areas outside the jurisdiction of the Russian Federation'. The full text of the consti-tution of the Republic of Tatarstan is available online at www.tatar.ru/constitution.html.

67 'Primet li Kreml Konstitutsiyu Tatarstana, kak ee Prinyal Nemtsov?', *Respublika Tatarstan*, 30 April 2002.

68 Ukaz Prezidenta Rossiiskoi Federatsii No. 1486 ot 10.08.00 'O dopolnitelnykh merakh po obespecheniyu edinogo pravovogo prostranstva Rossiiskoi Federatsii'. Online. Available at http://document.kremlin.ru/index.asp (accessed 20 April 2009).

69 Constitution of the Republic of Tatarstan, Article 21.

70 Ibid., Article 8.

71 'Prokuror Oprotestovyvaet Zakon "O Yazykakh Respubliki Tatarstan"', *Vremya i Dengi*, 23 May 2003.

72 'Genprokuratura Ishchet Pravdy', *Parlamentskaya Gazeta*, No. 934.

73 Constitution of the Republic of Bashkortostan, Article 87.25 and Article 100. On line. Available at http://constitution.garant.ru/DOC_17600023.htm#sub_para_N_603 (accessed 20 April 2009).

74 Ibid., Article 85.

75 Ibid., Article 87.25.

76 'Strakhi i Strasti Vlasti', *Nezavisimaya Gazeta*, 13 May 2003.

77 The full text of the law, available online at http://refugee.memo.ru/For_All/law.nsf/0/09478686826fb36ac3256bfd007a7b35?OpenDocument (accessed 20 April 2009).

78 'Stavropolsky Krai Budet Ustanavlivat Kvoty na Migratsiyu', *grani.ru*, 9 July 2003.
79 The leader of Chuvashiya Fedorov acted as a voice of gubernatorial resistance by appealing to Russia's Constitutional Court to render the presidential initiatives unconstitutional. Initially, Fedorov planned to mobilise the entire Federation Council and launch this case collectively. However, only two regional legislatures – those of Yakutiya and Adygeya – decided to back Fedorov's collective claim.
80 Postanovlenie Konstitutsionnogo Suda RF po delu proverki konstitutsionnosti otdelnykh polozhenii Federalnogo Zakona 'Ob obshchikh printsipakh organizatsii zakonodatelnykh i ispolnitelnykh organov vlasti subektov RF' v svyazi s zaprosom Gosudarstvennogo Sobraniya Respubliki Sakha i Soveta Respubliki Gosudarstvennogo Soveta Respubliki Adygeya, 04.04.02. Online. Available at the Official website of the Constitutional Court of the Russian Federation, http://ks.rfnet.ru/pos/p8_02.htm (accessed 20 April 2009).
81 In its original version, this provision theoretically created an opportunity for suspending the governors on the basis of various criminal matters. In particular, it does not stipulate the precise criteria in accordance with which the president should either approve or reject the suspension request made by the General Prosecutor. Therefore, interpretation of the severity of the criminal matter remained unclear. This situation changed in February 2007, when the criminal code was expanded to include *all* types of offences, not just 'severe' ones (see Gromov 2007).
82 Federalnyi Zakon FR 'O vnesenii izmenenii i dopolnenii v Federalnyi Zakon RF "Ob obshchikh printsipakh organizatsii zakonodatelnykh (predstavitelnykh) i ispolnitelnykh organov gosudarstvennoi vlasti subektov Rossiiskoi Federatsii"', Moscow, Kremlin, 4 July 2003. Online. Available at www.akdi.ru/gd/proekt/091202GD.shtm (accessed 20 April 2009).
83 Ibid.
84 Ibid.
85 'Uzdechka dlya Gubernatorov', *Nezavisimaya Gazeta*, 25 May 2007.
86 'Vybivanie Pokazatelei', *Kommersant Vlast*, 9 July 2007.
87 'Figurant-Gubernator', *Nezavisimaya Gazeta*, 17 May 2004.
88 'Kto Izbavlyaetsya ot Slabogo Zvena v Vertikali Vlasti', *Obshchaya Gazeta*, 24 May 2007.
89 'Regionalnoe Izmerenie: Lokomotivy Obshchego Dela', *Vedomosti*, 5 July 2005.
90 'Byudzhetnaya Vertikal. Gubernatorov Podchinili Premeru', *Gazeta*, 27 January 2006.
91 'Klyuch ot Gubernatorov', *Vremya Novostei*, 4 July 2005.
92 'Rospusk Vozmozhen', *Ekspert*, 6 June 2007.
93 Author's interview with A. Chirikova of the Institute of Sociology, Moscow (RAN).
94 'The Russian Elite Fears Integration with the West', *Vek*, No. 43, 1 November 2001.
95 Ukaz Prezidenta RF ot 13.05.00 No. 849 'O polnomochnom predstavitele Prezidenta Rossiiskoi Federatsii v Federalnom Okruge'. Online. Available at http://president.kremlin.ru/test_acm2/text/psmes/2000/05/31383.shtml (accessed 20 April 2009).
96 'Putin Podderzhal svoikh Polpredov Zhivoi Siloi', *Kommersant Daily*, 13 July 2000.
97 'Predstavlenie Nachinaetsya', *Kommersant Daily*, 20 May 2000.
98 Ukaz Prezidenta No. 849.
99 'Pro polpredov i lyudei', *Vremya Novostei*, 24 April 2006.
100 Author's interview with I.D. Zhuk, People's Deputy in the Moscow Regional Duma; see also 'Prava Cheloveka v Regionakh Rossiiskoi Federatsii', Belgorodskaya Oblast, Doklad 2002, *Prava Cheloveka v Rossii: Human Rights*. Online. Available at www.hro.org/docs/reps/2001/belg/2–1.htm (accessed 20 April 2009).
101 Author's Interview with I.D. Zhuk.
102 'Pro Polpredov i Lyudei', *Vremya Novostei*, 24 April 2006.

4 Institutions of centre-regional integration: monocentrism and its potential implications

1 Kremlin, http://president.kremlin.ru/STCWGroups.shtml.
2 'Putin Uspokoil Sovet Zakonodatelei', *Nezavisimaya Gazeta*, 19 February 2003.
3 'Perechen Federalnykh Tselevykh Programm i Federalnykh Programm Razvitiya Regionov na 2005 god', The Government of the Russian Federation. Online. Available at www.programs-gov.ru/cgi-bin/fcp_list.cgi (accessed 20 April 2009).
4 'Gossovet Budet Borotsya s Korruptsiei', *Nezavisimaya Gazeta*, 19 December 2001.
5 'Gubernatorov Budut Snimat Izyashchno', *gazeta.ru*, 21 November 2002. Online. Available at www.gazeta.ru/2002/11/21/gubernatorov.shtml (accessed 20 April 2009).
6 Postanovlenie Konstitutsionnogo Suda RF po delu proverki konstitutsionnosti otdelnykh polozhenii Federalnogo Zakona 'Ob obshchikh printsipakh organizatsii zakonodatelnykh i ispolnitelnykh organov vlasti subektov RF' v svyazi s zaprosom Gosudarstvennogo Sobraniya Respubliki Sakha i Soveta Respubliki Sakha, Soveta Respubliki Adygeya, 04 April 2002, *Sobranie Zakonodatelstva Rossiiskoi Federatsii*, No. 15, 15 April 2002, pp. 4057–76; Federalnyi Zakon FR 'O vnesenii izmenenii i dopolnenii v Federalnyi Zakon RF 'Ob obshchikh printsipakh organizatsii zakonodatelnykh (predstavitelnykh) and ispolnitelnykh organov gosudarstvennoi vlasti subektov Rossiiskoi Federatsii'', Kremlin, 4 July 2003, *Sobranie Zakonodatelstva Rossiiskoi Federatsii*, No. 27, 7 July 2003, pp. 6553–8.
7 The Government of the Russian Federation, 2005, http://council.gov.ru/sz/reshsz.htm.
8 Federalnyi Zakon 'O poryadke formirovaniya Soveta Federatsii Federalnogo Sobraniya Rossiiskoi Federatsii', Moscow, Kremlin, 7 August 2000, *Rossiiskaya Gazeta*, 8 August 2000.
9 The political compositions of regional parties are available at the official websites of regional legislatures. The full list of these sites is available online at www.gov.ru/main/regions/regioni-44.html, or www.regions.ru/cities/index.html.
10 'Za Rodnoi Ural, za Gubernatora Rosselya', *Nezavisimaya Gazeta*, 23 April 2002.
11 Official website of St. Petersburg Legislature, www.assembly.spb.ru. (accessed 20 April 2009).
12 'Bolshaya Partiinaya Mechta', *Vremya Novostei*, 4 March 2002.
13 'Delovaya Partiya', *Vedomosti*, 2 October 2008.
14 Central Electoral Commission of the Russian Federation. Online. Available at http://gd2003.cikrf.ru/WAY/76799135/sx/art/76805049/cp/1/br/76799124 (accessed 20 April 2009).
15 Itogi Vyborov v Gosdumu, Official site of the Russian State Duma, www.gduma.ru/itogi.htm.
16 'PR-Ukhod', *Vedomosti*, 30 August 2007.
17 'Chukotka dlya Vsekh', *Vedomosti*, 21 November 2003.
18 'U Putina Ryvok ne Vyshel', *gazeta.ru*, 26 November 2002. Online. Available at www.gazeta.ru/parliament/articles/8894.shtml (accessed 20 April 2009).
19 Talk by Alexandr Zhukov on *Svoboda Slova*, *NTV*, 16 January 2004 (unpublished).
20 Federalnyi Zakon RF 'O politicheskikh partiyakh', Kremlin, 11 July 2001, *Sobranie Zakonodatelstva Rossiiskoi Federatsii*, Vol. 28, 16 July 2001, pp. 5739–70.
21 This policy, as well as the reconstitution of gubernatorial elections, had no apparent connection to the problem of terrorism. Thus, a number of commentators (Colton *et al.*, 2005) observed that Putin's administration employed the Beslan attack as a means of expanding its political authority.
22 In many regions local financial and political elites formed 'spontaneous' parties during gubernatorial and legislative elections in order to secure executive offices and seats within the regional legislatures. These parties disintegrated shortly after the elections had taken place (Olshanskii 2000, at pp. 45–6).
23 Federalnyi Zakon RF 'O vyborakh Prezidenta Rossiiskoi Federatsii', Kremlin, 10

January 2003, *Sobranie Zakonodatelstva Rossiiskoi Federatsii*, No. 2, 13 January 2003, pp. 359–486.

24 Alexandr Veshnyakov, 'Ya Nameren i Dalshe Vozglavlyat Tsentrizbirkom', *Nezavisimaya Gazeta*, 25 November 2002.

25 Federalnyi Zakon RF 'Ob osnovnykh garantiyakh izbiratelnykh prav i prava na uchastie v referendume grazhdan Rossiiskoi Federatsii', Kremlin, 12 June 2002, *Sobranie Zakonodatelstva Rossiiskoi Federatsii*, No. 24, 17 June 2002, pp. 6074–213.

26 'Sekretar GenSoveta Partii Edinaya Rossiya Vyacheslav Volodin: Osnova Nashei Ideologii – Politika Prezidenta Putina', *Izvestiya*, 22 February 2007.

27 'Bez Litsa', *Vedomosti*, 20 August 2007.

28 Federalnyi Zakon RF 'O vnesenii izmenenii i dopolnenii v Federalnyi Zakon RF "Ob obshchikh printsipakh organizatsii zakonodatelnykh (predstavitelnykh) i ispolnitelnykh organov gosudarstvennoi vlasti subektov Rossiiskoi Federatsii"', Kremlin, 4 July 2003, *Sobranie Zakonodatelstva Rossiiskoi Federatsii*, No. 27, 7 July 2003 (Part 2), pp. 6553–80.

29 'Kazani – Pryanik, Ufe – Knut', *Nezavisimaya Gazeta*, 25 November 2002.

30 Ibid.

31 Ibid.

32 'Depresanty. Novaya Programma Prezhnego Kursa', *Sovetskaya Rossiya*, 11 August 2001.

33 'Regiony Trebuyut Zemli i Voli', *Vremya Novostei*, 9 September 2004.

34 Ibid.

35 'Shaimiev – Putin: Nichya s Prodolzheniem', *Nezavisimaya Gazeta*, 12 March 2002.

36 'Gref Pokusilsya na Svyatoe', *Vedomosti*, 8 June 2005.

37 www.government.gov.ru.

38 'Kreml Pognalsya za Dvumya Vertikalyami', *Kommersant Daily*, 5 October 2005.

39 'Zhenshchine ne Udobno Otkazat Spikeru', *Izvestiya*, 14 January 2002.

40 'Ten Skuratova', *Nezavisimaya Gazeta*, 16 January 2002.

41 'Ne Byvat Kokhu Senatorom', *Kommersant Daily*, 28 March 2002; 'Vtoraya Popytka Alfreda Kokha', *Nezavisimaya Gazeta*, 28 May 2002.

42 Federalnyi Zakon 'O vnesenii izmenenii i dopolnenii v Federalnyi Zakon "O poryadke formirovaniya Soveta Federatsii Federalnogo Sobraniya Rossiiskoi Federatsii" i Federalnyi Zakon "O statuse Chlena Soveta Federatsii i Statuse Deputata Gosudarstvennoi Dumy Federalnogo Sobraniya Rossiiskoi Federatsii"', 16 December 2004, *Sobranie Zakonodatelstva Rossiiskoi Federatsii* No. 51, 20 December 2004, pp. 11643–5; Federalnyi Zakon 'O vnesenii izmenenii v Federalnyi Zakon "O statuse chlena Soveta Federatsii i statuse deputata Gosudarstvennoi Dumy Federalnogo Sobraniya Rossiiskoi Federatsii"', *Sobranie Zakonodatelstva Rossiiskoi Federatsii*, 9 May 2005; see also Federalnyi Zakon 'O Poryadke Formirovaniya Soveta Federatsii Federalnogo Sobraniya Rossiiskoi Federatsii', 5 August 2000, *Sobranie Zakonodatelstva Rossiiskoi Federatsii*, No. 32, 7 August 2000, pp. 6247–51.

43 Federalnyi Zakon 'O vnesenii izmenenii i dopolnenii v Federalnyi Zakon "O poryadke formirovaniya Soveta Federatsii Federalnogo Sobraniya Rossiiskoi Federatsii" i Federalnyi Zakon "O Statuse Chlena Soveta Federatsii i Statuse Deputata Gosudarstvennoi Dumy Federalnogo Sobraniya Rossiiskoi Federatsii"', Kremlin, 16 December 2004, Vol. 51, 20 December, *Sobranie Zakonodatelstva Rossiiskoi Federatsii*, pp. 11643–5.

44 'Piterskaya Metla v Sovete Federatsii', *Nezavisimaya Gazeta*, 19 December 2001.

45 Putin Address to the Federal Assembly 16 April 2007, available at 'Poslanie Federalnomu Sobraniyu Rosiiskoi Federatsii Prezidenta Rossii Vladimira Putina. Polnyi Tekst, *Rossiiskaya Gazeta*, 27 April 2007.

46 'Desyat Let bez Prava na Sovfed', *Rossiiskaya Gazeta*, 9 June 2007.

47 *Kommersant Daily*, 5 December 2007.

48 'Aushev Says the Kremlin did not Push him Out', *The Moscow Times*, 25 January 2002.

49 'Vertikal Vlasti Protknula Sovet Federatsii', *gazeta.ru*, 31 January 2002, www.gazeta. ru/2002/01/30/vertikaljvla.shtml (accessed 20 April 2009).
50 'Koloda Rossiiskoi Federatsii', *Kommersant Vlast*, 1 December 2003.
51 *Svoboda Slova*, *NTV*, 30 May 2003 (unpublished).
52 'Ves Sovet Federatsii', *Kommersant Vlast*, No. 7, 26 February 2002.
53 'Tsena Senatorskogo Kresla. Lobbiruyutsya li v Sovete Federatsii Regionalnye Interesy?' *Saratovskaya GTRK*, 24 October 2002. Online. Available at http://saratov.rfn. ru/region/rnews.html?id=2125&rid=387 (accessed 20 April 2009).
54 Pryamaya Rech, 'Chto Vy Teper Budete delat?', *Kommersant Daily*, 2 October 2007.
55 'Smena Vetvi', *Vedomosti*, 8 October 2003.
56 'Parovozy Otseplyayutsya', *Vedomosti*, 5 December 2007.
57 'Edinaya Rossiya Poochshrila Gubernatorov za Vysokie Resultaty na Vyborakh', *Kommersant Daily*, 18 December 2007.
58 *RIA Novosti*, 21 February 2005.
59 'Vsego Odin Million', *Obshchaya Gazeta*, 30 September 2008.
60 'Mnogie Rossiiskie Partii Mogut Ischeznut posle Proverki Mingina', *Russia in the World*, 21 March 2005.
61 Federalnyi Zakon 'O vyborakh Deputatov Gosudarstvennoi Dumy Federalnogo Sobraniya Rosiiskoi Federatsii', enacted on 18 May 2005, *Rossiiskaya Gazeta*, 24 May 2005.
62 'Ushla v Istoriyu', *Ekspert*, 16 November 2007.
63 Online. Available at www.wciom.ru/novosti/reitingi/reitingi-gosudarstvennykh-institutov.html (accessed 1 October 2008).
64 'My ne poidem za mironovym, my poidem za Putinym', *Kommersant Daily*, 23 November 2007.
65 'Spravedlivaya Rossiya Privedet v Dumu Proverennye Kadry', *Kommersant Daily*, 6 December 2007.
66 'Gubernator – eto Nadolgo', *Vedomosti*, 21 February 2005; 'Krasnyi Poyas Raskalilsya Dobela', *Nezavisimaya Gazeta*, 21 March 2005.
67 Lijphart (1999, pp. 205–12, and in particular p. 212) measured the strength of bicameralism through 'symmetry' and 'incongruence', implying in the first case the method of selection and constitutional power redistribution and in the second case, the extent of electoral over-representation.
68 'Nikto ne Khotel Ustupat', *Kommersant Vlast*, No. 17 (386), 28 May 2000.
69 'Mneniya Predstavitelei Regionalnoi Administrativnoi Elity o Rezhime Vlasti', Russian State Service Academy, October 2001, www.rags.ru/s_center/opros/polit_ elita/index.htm.
70 'Plach Yaroslavlya', *Vedomosti*, 6 October 2005.
71 'Rossiya: Ee Nastoyashchee i Budushchee v Obshchestvennom Mnenii', Russian State Service Academy, 21–27 June 2005. Online. Available at www.rags.ru/content. php?id=280 (accessed 20 April 2009).
72 'Rossii Nedaleko do Kirgizii', *gazeta.ru*, 11 April 2005. Online. Available at www. gazeta.ru/2005/04/11/oa_154279.shtml (accessed 20 April 2009).

5 Business and politics in Russia's regions: the problem of political style

1 The 'oligarchic purge' targeted many prominent executives: the head of *Gazprom* Rem Vyakhirev, the minister of transportation Nikolai Aksenenko, the former owner of *Sibneft* Boris Berezovskii, media magnate Vladimir Gusinskii, the head of *Sibur* Yakov Golodovskii, former Prime Minister Viktor Chernomyrdin (who was sent into political exile as the Russian Federation's Ambassador to Ukraine), a number of *Yukos* executives and the head of the presidential administration Alexandr Voloshin, who quit his post voluntarily in protest at the attacks on big business during autumn 2003.
2 It is perhaps interesting to mention that Deripaska married the eldest daughter of

Yeltsin's son-in-law V. Yumashev. This step solidified the political-economic positions of the Yumashev–Yeltsin–Dyachenko–Abramovich clan.

3 'Semeinyi Monstr Zaglatyvaet Ekonomiku Rossii', *Moskovskii Komsomolets*, 23 December 2002; see also Orlov (2002).

4 Alexandr Uss, a representative of 'the family' group supported by *Russian Aluminium*, ran against the head of *Norilsk Nikel* Alexandr Khloponin. Khloponin won the race but the results were contested by the regional electoral committee, which was dominated by 'family-oriented' representatives of *Krasnoyarskii Aluminium*. Following Khloponin's complaint, on 3 October 2002 Putin signed a decree that appointed Khloponin as governor of Krasnoyarsk Krai. Uss subsequently retained his position as head of the regional assembly.

5 Besides extensive regional influence, 'the family' retained serious power connections at the centre until the end of 2003. Prime Minister Kasyanov, the head of the presidential administration Voloshin, the head of the territorial department of the presidential administration Surkov, and former head of the Security Council Rushailo represented the group's interests at the federal centre. It had become clear that Kasyanov had lobbied for an increase in the import quotas for second-hand foreign cars to support the industrial capacities of 'the family'-owned *GAZ* and *Ruspromavto* and pressed for a delay in Russia's accession to the WTO to protect 'the family'-owned insurance leaders *Ingosstrakh* and *Rosno*.

6 'Sovet Federatsii: Palata Reginov ili Biznes-Lobby?', *Stavropolskaya Pravda*, 6 April 2002.

7 'Gubernator za Bortom', *Nezavisimaya Gazeta*, 26 November 2002.

8 'Nenetskaya Neft Ostalas u Butova', *gazeta.ru*, 15 January 2001. Online. Available at www.gazeta.ru/2001/01/15/neneckaaneft.shtml (accessed 20 April 2009).

9 'Ne Luchshe li Doveritsya Izbiratelyu?', *Parlamentskaya Gazeta*, 14 December 2000.

10 'Irkutskie Vybory: Borba Ideologii ili Teatr Absurda', *Nezavisimaya Gazeta*, 14 August 2001.

11 'Glavoi Irkutskenergo Stal Alyuminshchik so Stazhem', *gazeta.ru*, 5 July 2001. Online. Available at www.gazeta.ru/2001/07/05/ener24551.shtml (accessed 20 April 2009).

12 'Gubernatora Magadanskoi Oblasti Nazovet lish Vtoroi Tur', *Nezavisimaya Gazeta*, 4 February 2003.

13 'Zima – na Moroz, Narod – na Vybory', *Rossiiskaya Gazeta*, 9 December 2003.

14 Ilya Yuzhanov interview, *Vedomosti*, 5 December 2002.

15 Data collected from official websites of regional administrative bodies, available online at www.kremlin.ru.

16 'Luchshie Lobbisty Rossii – Mai 2002', *Nezavisimaya Gazeta*, 26 June 2002.

17 'Sto Vedushchikh Politikov Rossii v Noyabre', *Nezavisimaya Gazeta*, 2 December 2002.

18 'Sto Vedushchikh Politikov Rossii v Oktyabre', *Nezavisimaya Gazeta*, 31 October 2003.

19 'Sto Vedushchikh Politikov Rossii v Noyabre', *Nezavisimaya Gazeta*, 1 December 2003.

20 Governor Nikolai Torlopov often admitted his friendly relations with the head of *SUEK* (a coal-mining branch of *MDM*) Oleg Miserva. Deputy governor Nikolai Levitskii was an ex-executive of *Evrokhim*, a branch of MDM.

21 'SUEK-Baikal Prishel v Komi', *Vedomosti*, 26 November 2002.

22 'Gubernator v Roli Obmanutogo Zhenikha', *Rossiiskaya Gazeta*, 20 August 2003.

23 'Oligarkh Gubernatoru ne Tovarishch?', *Nezavisimaya Gazeta*, 10 February 2004.

24 'Ostryi Zapakh Ammiaka: Arestovan Direktor Nevinomysskogo Azota', *Vremya-MN*, 6 July 2001.

25 'Kogda "Rostovskii Kotelshchik" Pokrasneet ot Styda?', *Versiya*, No. 20, 2003.

26 'Kazani – Pryanik, Ufe – Knut', *Nezavisimaya Gazeta*, 25 November 2002.

27 'Zakaz na Bratskii LPK', *Trud*, 26 December 2001.
28 'Za Gubernatora! Za Sibneft!', *Moskovskii Komsomolets*, 18 January 2001.
29 'Putin Prizval Biznes k Sotsyalnoi Otvetstvennosti', *Vedomosti*, 14 November 2003.
30 'Odnim Klyuchem Menshe', *Trud*, 1 October 2004.
31 Federalnyi Zakon 159-FZ ot 11.12.2004, *Rossiiskaya Gazeta*, 15 December 2004.
32 www.csr.gov.uk/feature.shtml (accessed 24 September 2007).
33 'Putin Prizval Oligarkhov k Sotsialnoi Otvetstvennosti', *Vedomosti*, 2004.
34 'Korporativnaya Sotsialnaya Otvetstvennost v Rossii: Teoriya i Praktika', *Vestnik Soveta Federatsii*, No. 26 (278), 2005, p. 16.
35 *Informatsionnaya Otkrytost Politiki Rossiiskikh Kompanii*, Mosow: Assotsiatsiya Menedzherov Rossii, 2004, p. 10.
36 *Korporativnaya Sotsialnaya Otvetstvennost: Obshchestvennye Ozhidaniya, Potrebiteli, Menedzhery, SMI i Chinovniki Otsenivayut Sotsialnuyu Rol Biznesa v Rossii*, Moscow: *Assotsiatsiya Menedzherov Rossii*, 2004, p. 13.
37 Ibid.
38 *Vestnik Soveta*, pp. 9–10.
39 'Valeriya Shantseva Peresazhivayut na Volgu', *Kommersant Daily*, 3 August 2005.
40 Tuleev's Budgetary Address to Regional Assembly, 27 October 2004. Online. Available at www.kemerovo.su/PRESS/Mess/Text/vistup.asp (accessed 24 September, 2007).
41 'Tuleev Obeshchaet "Amtelu" Problemy', *Vedomosti*, 5 May 2005.
42 Administration of Kemerovo Oblast Press Release, 13 July 2005. Online. Available at www.kuzbassinvest.ru/news.shtml?id=1121172017_9 (accessed 24 September 2007).
43 *Vestnik Soveta*, p. 12.
44 Alexandr Filipenko interview, 'Gubernator Filipenko: Segodnya Vazhnee Umet Prodat Chem Proizvesti', *Politicheskii Zhurnal*, No. 13, 11 April 2005.
45 *TNK-BP* Report 'Vneshnyaya Sotsialnaya Politika TNK-BP: ot Otdelnykh Proektov k Edinoi Strategii', Moscow February 2005, p. 7. Online. Available at www.tnk-bp.ru/social/external/ (accessed 24 September 2007).
46 'Nezamenimyi Abramovich', *Vedomosti*, 19 January 2007.
47 'Vekselberingovo More', *Vedomosti*, 11 August 2005.
48 Alexandr Veshnyakov Interview, 'Alexandr Veshnyakov Nadeetsya, chto Otmena Vyborov – Vremennaya Mera', *Vremya Novostei*, 10 February 2005.
49 Federalnyi Zakon No. 159-FZ, 11 December 2004, Moscow, *Rossiiskaya Gazeta*, 15 December 2004; Ukaz Prezidenta Rossiiskoi Federatsii No. 1603, 27 December 2004, *Rossiiskaya Gazeta*, 29 December 2004.
50 'Irkutskie Vybory: Borba Ideologii ili Teatr Absurda', *Nezavisimaya Gazeta*, 14 August 2001.
51 Kiril Tremasov, 'Vopros o Novom Irkutskom Gubernatore – Predmet Torga Krupnykh FPG', *Nezavisimaya Gazeta*, 4 August 2005.
52 'Gubernator s Bolshoi Dorogi', *Kommersant Daily*, 15 August 2005.
53 Standard and Poor's Analytical Report, 'Khanty-Mansiiskii Avtonomnyi Okrug – Kreditnyi Reiting', 23 March 2005. Online. Available at www.standardandpoors.ru/page.php?path=creditlist&pagenum=1 (accessed 24 September 2007).
54 'Neelov Ostanetsya na Yamale', *Vedomosti*, 18 February 2005.
55 Law 159-FZ.
56 'Regionalnye Elity Pereshli v Kontrataku', *Nezavisimaya Gazeta*, 17 March 2005.
57 'Put k Beregu Udachi', *Dalnevostochnyi Kapital*, 9 September 2001.
58 'Gubernator – v Kameru, Senator – v Otstavku', *Nezavisimaya Gazeta*, 24 May 2006.
59 'A Teper Lebed', *Vedomosti*, 25 May 2006.
60 'Zemletryasenie Razrushilo Nadezhdy Sakhalinskogo Gubernatora', *Kommersant Daily*, 8 August 2007.
61 Author's interview with regional Duma deputy Evgenii Main and the head of the regional *Interfax* bureau Evgenii Solovev, 20 July 2005 and 19 July 2005.

62 Author's interview, 22 July 2005.
63 'Vybor, Kotorogo Net', *Izvestiya Saratov*, 12 May 2005.
64 Alexei Motorkin interview, 'Politicheskii Krest Ekonomicheskoi Elity', *Afanasii-Birzha Rossiiskii Ekonomicheskii Ezhenedelnik*, 29 June 2005.
65 'Smena Konfiguratsii', *Permskii Obozrevatel*, 8 August 2005.
66 Alexandr Khloponin interview, 'Krai i Ploshchadka', *Expert*, 21 April 2003.
67 'Partiya Millionerov', *Vedomosti*, 22 October 2003.
68 'Edinaya Rossiya Reshila Podderzhat Dmitriya Zelenina na Vyborakh Gubernatora Tveri', *lenta.ru*, 21 October 2003. Online. Available at http://news.iof.ru/lenta.ru/2003/10/21/30 (accessed 20 April 2009).
69 Gubernatora Magadanskoi Oblasti Nazovet lish Vtoroi Tur', *Nezavisimaya Gazeta*, 4 February 2003.
70 See official website of the *Edinaya Rossiya* party and its regional membership pages; available online at www.edinros.ru, accessed 24 September 2007.
71 'Poterya Orientatsii', *Vedomosti*, 24 September 2007.
72 'Krasnoyarskii Krai Budut Lechit', *Vremya-MN*, 27 November 2002.
73 'Abramovich i Invalidy', *Vedomosti*, 18 December 2002.
74 'Khloponin i Deripaska Budut Igrat po Pravilam', *Izvestiya*, 15 November 2002.
75 'Biznes i Biznes Skhemy: Transfertnoe Tsenoobrazovanie Lezhit v Osnove Nashikh Bed', *Nezavisimaya Gazeta*, 26 June 2001.
76 'Tsena Neftyanogo Voprosa', *Kommersant Daily*, 21 May 2001; 'V Rossii tak Nelzya', *Vedomosti*, 4 October 2005.
77 Following the dissolution of the USSR, Baikonur became the territory of Kazakhstan. However, in 1995, in accordance with an agreement between the Russian and Kazakh governments, a Russian civilian administration was formed. It took responsibility for the industrial and residential areas of the town and its infrastructure. Since 1 January 1996, the town has officially been under Russian economic governance. The mayor of Baikonur is appointed by the president of the Russian Federation in agreement with the president of Kazakhstan.
78 'Oligarkhi Vyshli v Otkrytyi Kosmos', *Novaya Gazeta*, 18 March 2002.
79 'Orbitalnaya Skhema MNPZ: Nalogoviki Trebuyut s Zavoda $135mln', *Vedomosti*, 11 March 2005.
80 'Oligarkhi Vyshli v Otkrytyi Kosmos', *Novaya Gazeta*, 18 March 2002.
81 'Za Bashkirskie Skhemy Zaplatyat Ufimskie NPZ', *Kommersant Daily*, 26 January 2005.
82 'Nezakonnost Baikonurskikh Skhem Bolshe ne Budet Podvergatsya Somneniyu', *Vedomosti*, 5 August 2005.
83 'Schetnaya Palata Provedet Totalnuyu Proverku Vnutrennikh Ofshorov', *lenta.ru*, 20 May 2004. Online. Available at http://lenta.ru/economy/2004/05/20/benefits/ (accessed 24 September 2007).
84 Alexei Kudrin interview, 'Gubernator Abramovich Obmenyal Lgoty na Zoloto', *Kommersant Daily*, 20 November 2003.
85 Ibid.
86 Alexei Kudrin interview, 'Otstavka Voloshina Sovpala s Kontsom Epokhi Yeltsina', *Kommersant Daily*, 3 November 2003.
87 Ibid.
88 'Neftyanaya Korova Ustala', *Rossiiskaya Gazeta*, 2 September 2005.
89 'Transfertnye Tseny: Minfin Nastaivaet', *Ekonomika i Zhizn*, No. 10, 5 October 2005.
90 'Na Kogo Napravlen Udar: Transfertnye Tseny', *Ekonomika i Zhizn*, No. 11, 5 November 2005.
91 Prior to Khodorkovskii's arrest in October 2003, *Yukos*, through its parliamentary inter-faction *Energiya* group, had sabotaged all governmental attempts at clamping down on the system.
92 'Lobbisty v Gosdume', *Russkii Fokus*, 15 September 2003; 'Syrevaya Duma 2003', *Vedomosti*, 9 October 2003.

93 'Doklad ob Ekonomike Rossii', World Bank Publications. March 2005, No. 10. Online. Available at http://ns.worldbank.org.ru/files/rer/RER_10_rus.pdf (accessed 24 September 2007).
94 'Posledstviya Dela Yukosa: Luchshe Chem Kazhetsya', *Pravda*, 24 July 2005.
95 'Sibneft Doplatila v Byudzhet $125 mln', *Vedomosti*, 18 May 2005.
96 'V Tyumeni – Pozvoleno. TNK-BP Ispolzuyut Transfertnye Tseny', *Vedomosti*, 4 August 2005.
97 'U TNK-BP Otnimayut Lgoty', *Expert*, 11 October 2006.
98 Author's Interview with Alla Chirikova, 11 July 2005.
99 Author's Interview with Yaroslavl politicians, 18–22 July 2005.
100 'Smestit Tsentr Pribyli', *Ekspert-Volga*, No. 28, 30 October 2006.
101 'SUEK Perenosit Tsentr Tribyli v Irkutskuyu Oblast, *Vedomosti*, 23 July 2004.
102 'Val Gamburtseva Ushel za $7 mln', *Kommersant Daily*, 12 March 2001; The state-controlled *Rosneft* subsequently purchased *Severnaya Neft* from Vavilov for $600 million. See 'Vavilov Stal Bogache na $600 mln', *Vedomosti*, 13 February 2003.
103 Federalnyi Zakon 122-FZ 'O vnesenii izmenenii v zakonodatelnye akty RF i priznanie utrativshimi silu nekotorykh zakonodatelnykh aktov RF v svyazi s prinyatiem federalnykh zakonov "O vnesenii izmenenii i dopolnenii v Federalnyi Zakon 'Ob obshchikh printsipakh organizatsii zakonodatelnykh i ispolnitelnykh organov gosudarstvennoi vlasti subektov RF' i 'Ob Obshchikh printsipakh organizatsii mestnogo samoupravleniya v RF'"', *Sobranie Zakonodatelstva Rossiiskoi Federatsii*, Moscow: 2004, p. 2711.
104 'Oleg Mitvol Predlozhil Otozvat Chast Litsenzii Britanskikh Zolotodobytchikov', *Rossiiskaya Gazeta*, 29 November 2006.
105 '*Gazpromneft* Znachitelno Otklonilas ot Ustanovlennykh Pokazatelei po Obemam Uglevodorodnogo Syrya', *Rossiiskaya Gazeta*, 19 December 2006.
106 'Chinovniki Uporyadochili Nedropolzovanie', *Sibirskaya Neft*, 10 August 2006.
107 'Uchitel Trutnev', *Vremya Novostei*, 20 July 2006.
108 'Rosprirodnadzor Predlagaet Otozvat Litsenzii u Lukoila', *Vedomosti*, 13 October 2006.
109 'Mitvol Postradal iz-za 200 mln Untsii Zolota', *Ekspert*, 8 December 2006.
110 'Surgutneftegaz Dvigaet na Yakutiyu', *Surgutskaya Tribuna*, 4 November 2003; see also 'Yukos Gonyat iz Yakutii', *Vedomosti*, 2 January 2004.
111 'Gossovet i Minprirody Delyat Nedra', *Pravda*, 15 January 2003.
112 'Khozhdeniya Oligarkhov v Kreml', *gazeta.ru*, 19 February 2003. Online. Available at www.gazeta.ru/2003/02/19/hozdeniaolig.shtml (accessed 20 April 2009).
113 'Biznesmeny Priravnyali Koruptsiyu k Terrorizmu', *Vremya Novostei*, 20 February 2003.
114 'Pogloshchenie bez Razresheniya', *Vedomosti*, 22 February 2002.
115 'V Polzu Bogatykh', *Vedomosti*, 10 December 2002.
116 'Revolyutsiya ot RSPP', *Vedomosti*, 17 January 2003.
117 'Ekonomika bez Kontura', *Vedomosti*, 19 December 2002.
118 'Borba Klanov Rodila Klona', *Novaya Gazeta*, 31 July 2003.
119 'Putin Lishil Oligarkhov Neprikosnovennosti', *gazeta.ru*, 24 September 2007.
120 'On Takoi Odin', *Vedomosti*, 3 October 2007.
121 'Strana Vzyatok: Rossiya Dostigla Afrikanskogo Urovnya Korruptsii', *Vedomosti*, 19 October 2005; 'Rossiya Opustilas na 143 mesto', *Vedomosti*, 26 September 2007.
122 'Otkat v Rossii: V Strane Proizoshel Korruptsionnyi Perevorot', *Vedomosti*, 16 November 2005.
123 Ibid.
124 'Spasibo Germanu Grefu', *Vedomosti*, 16 November 2005.
125 Postanovlenie Pravitelstva No. 581, 5 August 2000. Online. Available at www.garweb.ru/project/spprinfo/documents/12020402/12020402.htm (accessed 24 September 2007).
126 Author's interview with Evgenii Main, 20 July 2005.

127 'Missiya Ekonomicheskogo Soveta', *Vremya Novostei*, No. 8 (11), 2004.
128 Author's interview with Oleg Vinogradov, 20 July 2005.
129 'Mezhdu Khlebom i Vodoi', *Kommersant Vlast*, 27 February 2006.
130 'Na Inteko Zhaluyutsya Deputatam', *Vedomosti*, 23 August 2005.
131 Assotsiatsiya Menedzherov Rossii, November 2004.

6 The unintended consequences of gubernatorial appointments in Russia

1 Wheare (quoted by Lijphart 1999, p. 190), for example, famously refused to call India a federal polity due to the establishment of the gubernatorial appointment system and the constitutional right of the central government to dismiss state governments and to replace them with direct rule from the centre for the purpose of dealing with grave emergencies.
2 'Vybory v Yakutii: Tretya Popytka Nikolaeva', *Parlamentskaya Gazeta*, 9 August 2001.
3 'Kreml Podderzhit Dzasokhova na Vyborakh Glavy Severnoi Osetii', *Nezavisimaya Gazeta*, 25 July 2001.
4 'Putin Naznachil Khloponina Gubernatorom', *gazeta.ru*, 3 October 2002. Online. Available at www.gazeta.ru/2002/10/03/putinnazna4i.shtml (accessed 20 April 2009).
5 'Kreml Podderzhit Dzasokhova na Vyborakh Glavy Severnoi Osetii', *Nezavisimaya Gazeta*, 25 July 2001.
6 This statement might initially be seen as an exaggeration. Indeed, given the federation's president right to nominate gubernatorial candidates, the parliaments cannot select an executive on the basis of inter-party bargaining or a party commanding a majority on the legislature's floor. Moreover, regional executives have retained the right to form their cabinets without parliamentary approval.
7 By the time of the reform, the only region that had never had a directly elected executive was Dagestan. Instead, it had a collective presidency named the State Council, composed of the representatives of the republic's ethnic groups and elected by the Constitutional Assembly, composed of the republic's legislatures and the delegates of municipal assemblies (Golosov 2003, p. 64).
8 This thesis was first introduced by Maurice Duverger (1964, pp. 217–26). It was subsequently deployed by many scholars (see Lijphart 1984, p. 165; Shugart and Carey 1992, pp. 206–8; Powell 1982, pp. 82–3).
9 Federalnyi Zakon 'Ob osnovnykh garantiyakh izbiratelnykh prav i prava na uchastie v referendume grazhdan Rossiiskoi Federatsii', Moscow, *Rossiiskaya Gazeta*, 2 February 2007.
10 The theoretical literature classifies this system as a subdivision of proportional representation (Lijphart 1986, pp. 145–8).
11 Ukaz Prezidenta Rossiiskoi Federatsii No. 1603 'O poryadke rassmotreniya kandidatur na dolzhnost vysshego dolzhnostnogo litsa (rukovoditelya vysshego ispolnitelnogo organa gosudarstvennoi vlasti) subekta Rossiiskoi Federatsii', 27 December 2004, *Rossiiskaya Gazeta*, 29 December 2004.
12 'President Budet Vybirat v Obstanovke Sekretnosti', *Nezavisimaya Gazeta*, 17 January 2005.
13 'V Ocheredi k Izbiratelyu No. 1', *Kommersant Vlast*, No. 6 (609), 14 February 2005.
14 'Men Nezavisimosti', *gazeta.ru*, 18 November 2005.
15 Of the 27 meetings Putin held with regional governors between January and May 2005, 11 resulted in a request of confidence and subsequent reappointment. 'Kremlevskie Vstrechi s Gubernatorskimi Posledstviyami', *Kommersant Daily*, 13 May 2005.
16 'Titov Poprosil Doveriya', *Vedomosti*, 21 April 2005.
17 'Dmitrii Medvedev Sovmeshchaet Priyatnoe s Sekretnym', *Kommersant Daily*, 8 April 2005.
18 'Sobyanin Prosit Doveriya', *Vedomosti*, 4 February 2005.
19 'Izderzhki Vertikali', *Profil*, No. 7, 28 February 2005.

20 'Gubernator – Eto Nadolgo', *Vedomosti*, 21 February 2005.
21 'Shantsev Otsidel Sto Dnei a Kirienko Vyshel na Svobodu', *Nezavisimoe Obozrenie, Nizhnii Novgorod*, 15 November 2005.
22 Federalnyi Zakon No. 159-FZ .
23 'Pervyi Atomnyi', *gazeta.ru*, 3 March 2005.
24 'Neelov Ostanetsya na Yamale', *Vedomosti*, 18 February 2005.
25 'Glavu Tuvy Okruzhayut Konkurentami', *Kommersant Daily*, 7 February 2007.
26 Ukaz Prezidenta Rossiiskoi Federatsii No. 1603.
27 'Kandidatov v Samarskie Gubernatory Stanovitsya vse Bolshe', *Kommersant Daily*, 5 April 2005; 'Titov Poprosil Doveriya', *Vedomosti*, 21 April 2005.
28 'U Putina net Drugikh Gubernatorov', *gazeta.ru*, 24 February 2005.
29 'Loyalnost v Mode', *Vedomosti*, 1 April 2005.
30 *Nezavisimyi Institut Vyborov* at www.vibory.ru. NP – the number of deputies that were not present during the voting.
31 'Vyshel iz Doveriya', *Vedomosti*, 1 June 2007.
32 'Titov Zadumalsya ob Otstavke', *Vedomosti*, 9 August 2007.
33 'Permskii Izlom', *Expert*, 2 August 2007.
34 'Kandidatura Gubernatora Opredelena', *Kommersant Daily*, 30 May 2007.
35 For example, plenipotentiary representative to the Ural Federal District Petr Latyshev openly declared that by enacting the new appointment system the centre did not intend to embark on reshuffling the composition of regional elites. See 'Petr Latyshev: Smena Regionalnykh Elit – eto uzhe Revolyutsiya', *Nezavisimaya Gazeta*, 28 April 2005.
36 'Chelovek iz *Yukosa* Ostalsya v Dolzhnosti', *gazeta.ru*, 3 March 2005. Zolotarev helped the nomination of *Yukos*'s ex-executive Vasillii Shakhnovskii to the Federation Council in order to grant Shakhnovskii the parliamentary immunity which could prevent his prosecution. Similarly, it has become known that Mikhail Khodorkovskii planned to become a Federation Council deputy from Evenkiya a few weeks before his arrest.
37 President Shaimiev interview with Nikolai Svanidze, *Zerkalo* programme (unpublished).
38 'Neelov Ostanetsya na Yamale', *Vedomosti*, 18 February 2005 and 'Filipenko Ostanetsya', *Vedomosti*, 8 February 2005.
39 'Mukhu Bei', *gazeta.ru*, 20 February 2006.
40 'Gubernatory Starozhily', *gazeta.ru*, 25 April 2005. Online. Available at www.gazeta.ru/2005/04/25/oa_155826.shtml (accessed 20 April 2009).
41 Source: official websites of regional administrations at www.kremlin.ru.
42 The region was liquidated in March 2008 due to the merger with Chita Oblast. Zhamsuev was subsequently transferred to Moscow to the post of an advisor to the head of the presidential administration Sergey Sobyanin.
43 'Mikhail Evdokimov za God Sobral Protiv Sebya Anshlag', *Vremya Novostei*, 20 March 2005.
44 'Mikhail Lapshin Otbilsya ot Impichmenta', *Kommersant Daily*, 1 April 2005.
45 'Regionalnye Elity Pereshli v Kontrataku', *Nezavisimaya Gazeta*, 17 March 2005.
46 'Dagestanskie SMI o Smene Vlasti v Respublike', *Regnum*, 27 February 2006, available at www.regnum.ru/news/597170.html.
47 The 1995–9 electoral cycle scored 7.3 due to a number of outliers such as Krasnoyarsk Krai, where as many as 40 parties formed 25 electoral blocs in the 1997 regional legislative election, and Moscow Federal city, in which 29 parties belonging to 20 blocs contested 35 seats in the city Soviet (Golosov 2003, p. 75).
48 *Nezavisimyi Institut Vyborov* at www.vibory.ru.
49 Author's interviews with Alexei Zudin, Natalya Zubarevich, and Yakov Pappe, July–August 2005.
50 Chukotka, with its two parties in parliament, represents an outlier. However, given Chukotka's unusual political situation dominated by the governorship of Russia's

richest man, Roman Abramovich, and his personal financial contributions towards the region's wealth, I excluded this region from the main count. Similarly, I have not included the Tyva legislature in the count, given the exceptionally difficult political relations within the region dominated by the struggle between two competing elite clans. The effective number of parliamentary parties falls to 4.3, if the Chukotka and Tyva two-party parliaments are taken into account.

51 www.vibory.ru/elects/reg-zak_r_05.htm#YaNAO.
52 'Voronezh ne Stal Stalingradom', *Vedomosti*, 22 March 2005.
53 'Krasnyi Poyas Raskalilsya Dobela', *Nezavisimaya Gazeta*, 21 March 2005.
54 'Chelyabinskie Demokraty Obedinyayutsya po Moskovskoi Skheme', *Kommersant Daily*, 4 October 2005.
55 'Edinyi Den Regionalnykh Vyborov', *Vremya Novostei*, 13 March 2006.
56 'Vosem Parlamentov Vyberut po Edinomu Stsenariyu', *Kommersant Vlast*, 10 March 2006.
57 'Skromnoe Obayanie Zaksobranii', *Nezavisimaya Gazeta*, 7 October 2004.
58 'Pravye Reshili Uiti Podalshe ot Kremlya', *Kommersant Daily*, 26 December 2005.
59 Party's Official Electoral Report, www.rodina.ru/otchet/regvybory.
60 'Vybory na Grani Vozmozhnogo', *Nezavisimaya Gazeta*, 13 March 2006.
61 'Partiya Pensionerov Vzyala Magadan', *Kommersant Daily*, 23 May 2005.
62 'Voskresnyi Anshlyus Edinoi Rossii', *gazeta.ru*, 11 March 2006.
63 'Vybory na Grani Vozmozhnogo', *Nezavisimaya Gazeta*, 13 March 2006.
64 'Povtoritsya li Mart v Dekabre?', *Rossiiskaya Gazeta*, 24 March 2007.
65 'Uslovnaya Pobeda: Edinoi Rossii Nuzhen Sparring-partner', *Vedomosti*, 14 March 2006.
66 Alexandr Kynev, 'Regiony Pridavlennye Vertikalyu', 5 February 2008. Online. Available at www.ryzhkov/publications/php?id=7817 (accessed 30 December 2008).
67 'Gubernator Peterburga Izbavit Gorod ot Provintsialnykh Kompleksov', *Izvestiya*, 1 April 2005.
68 'Yaroslavskie Deputaty Otkazalis Naznachat Gubernatora', *Izvestiya*, 22 April 2005.
69 Shaimiev *Zerkalo* interview.
70 'Irkutskie Deputaty Razocharovalis v Novom Gubernatore', *Kommersant Daily*, 23 September 2005.
71 'Uslovnaya Pobeda. Edinoi Rossii Nuzhen Sparring-partner', *Vedomosti*, 14 March 2006.
72 'Nizhegorodskie Kommunisty Mitinguyut', *Nezavisimaya Gazeta*, 4 March 2006.
73 'U Permskogo Kraya Poyavilsya Novyi Gubernator', *Kommersant Daily*, 11 October 2005.
74 'Ocherednaya Aktsiya Protesta v Preddverii Perenaznacheniya Eduarda Rosselya', *Obshchaya Gazeta Ural*, 25 November 2005.
75 'Vybory s Izyuminkoi', *Nezavisimaya Gazeta*, 5 November 2004.
76 Federalnyi Zakon No. 159-FZ, *Rossiiskaya Gazeta*, 25 July 2008.
77 '*Polpredy* Uglubilis v Kadrovyi Vopros', *Nezavisimaya Gazeta*, 13 January 2005.
78 *RLRFE* Report, 15 July 2005.
79 Nizhegorodskomu Gubernatoru Predlozhili Uiti Po-khoroshemu', *Kommersant Daily*, 25 April 2005.
80 'Nizhegorodskii Parlament Utverdil Shantseva', *Vedomosti*, 8 August 2005.
81 'Mikhail Men Prorabotaet v Ivanovo po-Luzhkovski', *Kommersant Daily*, 30 September 2005.
82 'Mikhail Lapshin Trebuet Ukhoda', *Kommersant Daily*, 5 October 2005.
83 'Regionalnye Elity'.
84 'Murtazu – v Otstavku, 'Boing' – k Podezdu', *Nezavisimaya Gazeta*, 7 April 2005.
85 'Obedinennyi Kamchatskii Krai Vozglavit Vitse-Gubernator Kamchatki', *Kommersant Daily*, 18 May 2007.
86 'Gubernatoriskie Rotatsii', *Moskovskie Novosti*, 7 September 2007.

7 Inter-regional relations and territorial problems

1 Berezovskii, 'Open Letter to the President'.
2 Ukaz Prezidenta RF ot 13 May 2000 No. 849 'O polnomochnom predstavitele Prezidenta Rossiiskoi Federatsii v Federalnom Okruge'.
3 Author's interview with Anatolii Alexandrovich Murashov, Mayor of Mytishchi, Moskva Oblast.
4 Working Documents of Analytical Groups of Moscow Regional Duma (unpublished).
5 'Vladimir Putin Poekhal Delit Norilsk', *Kommersant Daily*, 22 March 2002.
6 'Region Plus Region, a Chto v Minuse?', *Nezavisimaya Gazeta*, 17 June 2002.
7 Author's Interview with N.I. Rebchenko.
8 'Poslushayutsya li Oligarkhi RSPP?', *Vedomosti*, 9 December 2002.
9 'Bankiry Skinutsya na Lobbistov', *Vedomosti*, 23 December 2002.
10 'Druzhit Domami', *Vedomosti*, 30 December 2002.
11 'Oligarkhi Snova Obedinyayutsya', *Nezavisimaya Gazeta*, 22 December 2003.
12 Press Service of Plenipotentiary Representative in the Volga Federal District, available at www.pfo.ru, 22 November 2000.
13 Viktor Cherkesov interview, *Moskovskii Komsomolets*, 25 May 2001.
14 Author's interview with N.I. Rebchenko.
15 Poltavchenko interview, 'My Bystro Stareem. I Eto Problema', *Parlamentskaya Gazeta*, 9 September 2000.
16 Poltavchenko interview, 'Bez Rossii Rukhnet Mir', *Literaturnaya Gazeta*, 25 February 2003.
17 'Obsuzhdeny Ekonomicheskie Programmy Regionov Tsentra Rossii', *regions.ru*, 10 February 2001.
18 Ibid.
19 'Sovet Tsentralnogo Federalnogo Okruga Provel Vstrechu s Rukovoditelyami Prigranichnykh Gosudarstv', *regions.ru*, 23 April 2002.
20 Ibid.
21 Georgii Poltavchenko interview, 'Moskvu Razvernuli Litsom k Okrugu', *Rossiiskaya Gazeta*, 18 July 2002.
22 Poltavchenko interview, 'Bez Rossii Rukhnet Mir'.
23 Leonid Drachevskii interview, *Kommersant Daily*, 30 September 2000.
24 PRIME-TASS, 6 November 2001.
25 'Prezident Kyrgyzstana Akaev Vstretilsya s Polnomochnym Predstavitelem Prezidenta Rossii v Sibirskom Federalnom Okruge Drachevskim', *Obshchestvennoe Mnenie*, 1 October 2002.
26 'Apparat Polpreda v Yuzhnom FO', *Izvestiya*, 19 June 2001.
27 Author's interview with K.K. Khramov, 25 September 2002.
28 'Sibirskii Federalnyi Okrug Igraet Vazhneishuyu Rol v Razvitii Sotrudnichestva Rossii i Mongolii', *Delovoi Novosibirsk*, 3 July 2003.
29 'Mezhregionalnaya Asotsyatsiya "Sibirskoe Soglashenie" i Sibirskii Federalnyi Okrug Schitayut Neobkhodimym Postroit AES v Regione', *Delovoi Novosibirsk*, 3 March 2003.
30 Official Site of the Federation of Independent Trade Unions of Volga Federal District. Online. Available at www.fnpr-pfo.ru/ (accessed 20 April 2009).
31 'Bednye Rodstvenniki', *Novye Izvestiya*, 15 March 2007.
32 'Investory Predpochitayut Razvitye Regiony', *Vedomosti*, 4 October 2006.
33 'Drugaya Rossiya', *Novye Izvestiya*, 2 February 2007.
34 'Moskva Teper Provintsii po Nravu', *Vechernyaya Moskva*, 7 September 2006.
35 'Provintsialy Stali Terpimee k Moskvicham', *Izvestiya*, 5 September 2006.
36 'Bednye Rodstvenniki', *Novye Izvestiya*, 15 March 2007.
37 'Veter Peremen', *Novye Izvestiya*, 1 March 2007.
38 Ibid.
39 'Putin Nazval Ogromnoi Problemoi Bezrabotitsu Yuga Rossii', *Izvestiya*, 27 September 2005.

40 'Yuzhnyi Krest', *Vremya Novostei*, 26 September 2005.
41 'Strashnaya Bezopasnost', *Vremya Novostei*, 27 November 2007.
42 'Prodolzhenie Sleduet', *Vremya Novostei*, 15 October 2007.
43 'Chislo Vmesto Umeniya', *Vedomosti*, 14 September 2007; 'Za Zakrytymi Dveryami', *Rossiiskaya Gazeta*, 16 July 2005.
44 'Yuzhnyi Krest', *Vremya Novostei*, 26 September 2005.
45 'Yakutiya – Tot Eshche Subekt', *Novaya Gazeta*, 7 September 2006.
46 'Rvetsya Tam, Gde Tonko', *Agenstvo Politicheskikh Novostei*, 3 October 2006. Online. Available at www.apn.ru/publications/article10513.htm (accessed 20 April 2009).
47 'Dvukh Partii Malo', *Vedomosti*, 28 February 2008.
48 'Novyi Front', *Ekspert*, 3 October 2007; 'Fradkov Vozvrashchaetsya', *Rossiiskaya Gazeta*, 2 August 2007.
49 'Ovladet Vostokom', *Vedmosoti*, 6 August 2007.
50 'Ostrov Russkiy Predstavyat v Kannakh', *Rossiiskaya Gazeta*, 30 January 2008.
51 'Gosudarstvo Planiruet Vpyatero Uvelichit Finansovuyu Podderzhku Yuga Rossii', *Rossiiskaya Gazeta*, 13 July 2007.
52 'Glavy Subektov Sokhranili za Soboi Pravo Nazyvatsya Prezidentami', *Kommersant Daily*, 15 February 2001; 'Gosduma Ostavila Pravo za Glavami Respublik Naimenovatsya Prezidentami', *strana.ru*, 18 January 2002. Online. Available at www.strana.ru/news/103366.html (accessed 20 April 2009); 'Gosduma Pozvolila Glavam Respublik Ostatsya Prezidentami', *lenta.ru*, 17 June 2005. Online. Available at http://lenta.ru/news/2005/06/17/president/ (accessed 20 April 2009).
53 Speaking of Asian, African, and Caribbean states, Horowitz (1985, pp. 654–7) notes that, although the policies of ethnic preferences are universally regarded as exceptions and temporary expedients, many ethnically divided societies adopt them. Similar examples exist in the Western world. In Belgium, fixed language quotas are equally applied to the top of the federal and Brussels' regional services, judiciary, army, and diplomacy (Swenden 2002, p. 77.)
54 Source: Ministry of Economic Development of the Russian Federation, October 2004. 'O Tekushchei Situatsii v Ekonomike Rossiiskoi Federatsii v Yanvare-Sentyabre 2004 goda i Otsenkakh do Kontsa Goda'. Online. Available at www.budgetrf.ru/publications/2004 (accessed 20 April 2009).
55 Federalnyi Konstitutsionnyi Zakon ot 17 December 2001 'O Poryadke prinyatiya v Rossiiskuyu Federatsiyu i obrazovaniya v ee sostave novogo subekta Rossiiskoi Federatsii', *Rossiiskaya Gazeta*, 19 December 2001.
56 'Gubernatory Delyat Norilsk', *Nezavisimaya Gazeta*, 9 April 2002.
57 'V Blizhaishem Budushchem Ukrupneniya Regionov ne Budet', *Parlamentskaya Gazeta*, 23 September 2003; Viktor Cherkesov interview, 'Izmenitsya li Karta Strany?', *Argumenty i Fakty*, 26 December 2001.
58 'Ekonomika Vazhnee Etnicheskikh Avtonomii', *Vremya Novostei*, 17 April 2006.
59 'Doprosy bez Vykhodnykh', *Nezavisimaya Gazeta*, 3 May 2006; 'Prezident Nachal Uvolnyat Izbrannykh Gubernatorov', *Vremya Novostei*, 10 March 2005.
60 'V Razrabotke – Protsess Obedineniya Regionov', *Zhizn Regionov*, 21 May 2004.
61 'Lyubitel Krupnykh Form', *Vedomosti*, 7 February 2007.
62 'Khloponin Vystavil Schet', *Vedomosti*, 18 June 2004; see also Ukaz Prezidenta RF No. 412 'O merakh po sotsyalnomu razvitiyu Krasnoyarskogo Kraya, Taimyrskogo (Dolgano-Nenetskogo Avtonomnog Okruga) i Evenkiiskogo Avtonomnogo Okruga', 12 April 2005, Moscow, Kremlin. Online. Available at http://document.kremlin.ru/doc.asp?ID=027250 (accessed 20 April 2009).
63 Author interview with Natalya Zubarevich.
64 Oleg Kozhemyako Interview, 'Referendum po Obedineniyu Dolzhen Proiti s Polozhitelnym Rezultatom', *regnum*, 30 May 2005.
65 'Irkutsk i Ust-Orda Dogovorilis', *Nezavisimaya Gazeta*, 9 April 2004.

66 'Rosnefti Meshayut Kupit Val Gamburtseva', *gazeta.ru*, 24 January 2003. Online. Available at www.gazeta.ru/2003/01/24/rosneftjisev.shtml (accessed 20 April 2009).
67 'Neft Poidet Drugim Putem', *Izvestiya*, 9 December 2003.
68 'Putin Agitiruet za Khloponina', *Vedomosti*, 13 April 2005.
69 'Federalno-Korporativnoe Sliyanie', *Russkii Zhurnal*, 21 September 2006.
70 The governor of St Petersburg Valentina Matvienko was particularly keen on this idea. 'Iz Sankt-Peterburga v Super-Peterburg', *Nezavisimaya Gazeta*, 30 October 2006; also see 'Anatolii Lisitsyn: Silnye Dolzhny Obedinyatsya so Slabymi', *Nezavisimaya Gazeta*, 4 December 2002.
71 Alexandr Khloponin Interview, 'Eto Vtoraya Industrializatsiya Sibiri', *Vedomosti*, 16 March 2005.
72 'Lyubitel Krupnykh Form', *Vedomosti*, 5 February 2007.
73 'Evenkiya v Kommunnu ne Speshit', *Nezavisimaya Gazeta*, 11 April 2002.
74 'Obedinyai i Vlastvui', *Rodnaya Gazeta*, 4 June 2003.
75 'Protivniki Ukrupneniya Peredumali', *Nezavisimaya Gazeta*, 10 March 2004.
76 For nationality structure data see the official website of the Russian Government at www.gov.ru; see also Brubaker 1996, p. 36.
77 'Prezident Buryatii Shchitaet Narezku Federalnykh Okrugov Nepravilnoi', *Nezavisimaya Gazeta*, 18 March 2002.
78 Constitution of the Republic of Tatarstan. Online. Available at www.tatar.ru/?node_id=222 (accessed 20 April 2009).
79 'Tatarstan Vstroili v Vertikal', *Nezavisimaya Gazeta*, 20 November 2006. The Federation Council has subsequently blocked the Treaty, and many analysts have rightly seen the Kremlin's hand behind this. Indeed, on the verge of the 2007–8 electoral cycle, Putin could not have publicly denied the Republic its constitutional right to the Treaty and, therefore, entrusted this function to the head of the upper chamber Sergei Mironov. See also 'U Mintimera Shaimieva Otbirayut Podarok', *Nezavisimaya Gazeta*, 21 February 2007; 'Vtoroe Vzyatie Kazani', *Nezavisimaya Gazeta*, 19 December 2006.
80 Constitution of the Republic of Tatarstan.
81 'Derzhavnyi Orel Ne Prizhilsya na Severe', *Nezavisimaya Gazeta*, 3 July 2001; 'Federalnye i Okruzhnye Zakony Perevodyat na Yazyki Narodov Severa', *Sever-Press*, 21 November 2006.
82 Zakon No. 166-FZ 'O vnesenii dopolneniya v statyu 3 zakona Rossiiskoi Federatsii "O yazykakh narodov Rossiiskoi Federatsii" ot 11 December 2002', *Rossiiskaya Gazeta*, 14 December 2002; also see Zakon RF 'O yazykzkh narodov Rossiiskoi Federatsii ot 25 October 1991', *Sobranie Zakonodatelstva Rossiiskoi Federatsii*, No. 31, 1998, p. 3804.
83 Postanovlenie Konstitutsionnogo Suda Rossiiskoi Federatsii ot 16 November 2004 No. 16-P, *Rossiiskaya Gazeta*, 23 November 2004.
84 Federalnyi Zakon RF ot 1 June 2005 No. 53-FZ 'O gosudarstvennom yazyke Rossiiskoi Federatsii', *Rossiiskaya Gazeta*, 7 June 2005.
85 'Derzhavnyi Orel Ne Prizhilsya na Severe', *Nezavisimaya Gazeta*, 3 July 2001.
86 Doklad Goskomstata Rossii 'Ob Itogakh Vserossiiskoi Perepisi Naseleniya 2002 Goda' na Zasedanii Pravitelstva Rossiiskoi Federatsii 12 Fevralya 2004 goda, available at the official website of Russian Statistics Committee www.gks.ru/perepis/osn_itog.html; see also www.perepis2002.ru/index.html?id=17.
87 Ibid.

Bibliography

Adorno, T.W., Aron, B., Levinson, M.H., and Morrow W. (1950) *The Authoritarian Personality*, New York: Harper.

Agaptsov, S. (2001) 'Kak Razreshit Konflikt Dvukch Urovnei?', *Rossiiskaya Federatsiya Segodnya*, 13: 1.

—— (2003) 'Federalnyi Byudzhet 2004 i Byudzhetnyi Federalizm', *Sovet Federatsii*, 10: 14.

Agarwal, P. (1959) *The System of Grants-in-Aid in India*, London: Asia Publishing House.

Alexander, J. (2000) *Political Culture in Post-Communist Russia*, London: Macmillan Press.

Alexeev, M. (2000) 'The Unintended Consequences of Anti-Federalist Centralisation in Russia', *Program on New Approaches to Russian Security (PONARS) Memo Series*, No. 117. Online. Available at www.csis.org/ruseura/PONARS/policymemos/pm_0117.pdf (accessed 20 April 2009).

—— (1999) *Centre–Periphery Conflict in Post-Soviet Russia: A Federation Imperilled*, Basingstoke: Macmillan.

Alexeeva, L. (2007) 'Grazhdanskoe Obshchestvo: Let Cherez 10–15', *Vedomosti*, 20 February.

Ali, M. (1996) *Politics of Federalism in Pakistan*, Karachi: Royal Book Company.

Alm, J., Martinez-Vazquez, J., and Indrawati, S. (2004) *Reforming Intergovernmental Fiscal Relations and Rebuilding of Indonesia*, Cheltenham: Edward Elgar.

Almond, G. and Verba, S. (eds) (1963) *The Civic Culture: Political Attitudes and Democracy in Five Nations*, London: Sage Publications.

Arinin, A. (1999a) *Rossiiskii Federalism: Istoki, Problemy i Perspektivy Razvitiya*, Moscow: Soyuz.

—— (1999b) *Rossiiskii Federalism i Grazhdanskoe Obshchestvo*, Moscow: Izdanie Gosudarstvennoi Dumy.

—— (2000) *K Novoi Strategii Razvitiya Rossii. Federalizm i Grazhdanskoe Obshchestvo*, Moscow: Sever-Print.

Arinin, A. and Marchenko, G. (1999) *Uroki i Problemy Stanovleniya Rossiiskogo Federalizma*, Moscow: Inteltekh.

Aydingun A. and Aydingun I. (2004) 'The Role of Language in the Formation of Turkish National Identity and Turkishness', *Nationalism and Ethnic Politics*, 10(1): 415–32.

Badovskii, D., (2001) 'Sistema Federalnykh Okrugov i Institut Polnomochnykh Predstavitelei Prezidenta RF: Sovremennoe Sostoyanie i Problemy Razvitiya', Scientific-Research Institute of Social Systems of Moscow State University, Seriya *Nauchnye Doklady*, 7: 5.

Badovskii, D. and Shutov, A. (1995) 'Regionalnye Elity v Postsovetskoi Rossii: Osobennosti Politicheskogo Uchastiya', *Kentavr*, 6: 3–23.

Baker, W. and McCoy, C. (1991) *Fountainhead of Federalism: Heinrich Bullinger and the Covenantal Tradition*, Westminster: John Knox Press.

Bakvis, H. and Chandler, W. (eds) (1987) *Federalism and the Role of the State*, Toronto: University of Toronto Press.

Barnes, A. (2003) 'Russia's New Business Groups and State Power', *Post-Soviet Affairs*, 19(2): 154–86.

Barney, J. (1991) 'Firm Resources and Sustained Competitive Advantage', *Journal of Management*, 17: 99–120.

Bartsis, I. (2001) 'Federalnoe i Regionalnoe Zakonodatelstvo: Trebovanie Sootvetstviya', *Pravo i Politika*, No. 3.

Bednar, J. (2004) 'Judicial Predictability and Federal Stability Strategic Consequence of Institutional Imperfection', *Journal of Theoretical Politics*, 16(4): 432–46.

—— (2005) 'Federalism as a Public Good', *Constitutional Political Economy*, 16(2): 189–205.

Beer, S. (1993) *To Make a Nation: the Rediscovery of American Federalism*, Cambridge MA: Belknap Press.

Beesley, M. and Evans, T. (1978) *Corporate Social Responsibility*, London: Croom Helm.

Belousov, A. (2002) 'Kachestvo Rosta Rezko Snizhaetsya', *gazeta.ru*, 15 October. Online. Available at www.gazeta.ru/2002/10/12/andrejbelous.shtml (accessed 20 April 2009).

Benrabah, M. (2004) 'Language and Politics in Algeria', *Nationalism and Ethnic Politics*, 10(1): 59–78.

Berdyaev, N. (1955) *Istoki i Smysl Russkogo Kommunizma*, Paris: YMCA Press.

Berezovskii, B.A. (2000) 'Open Letter to the President of the Russian Federation', *Kommersant-Daily*, 30 May.

Berlin, I. Kelly, A., and Hardy, H. (1978) *Russian Thinkers*, London: Penguin.

Birch, A.H. (1957) *Federalism, Finance, and Social Legislation in Canada, Australia, and the United States*, Oxford: Oxford University Press.

Blattberg, C. (2006) 'Secular Nationhood? The Importance of Language in the Life of Nations', *Nations and Nationalism*, 12(4): 597–612.

Borisov, I. (2004) 'Razvitie Mnogopartiinosti v Rossii', *Nezavisimyi Institut Vyborov*. Online. Available at www.vibory.ru/discussion/borisov-p.htm (accessed 2 November 2006).

Borodai, O. (2002) 'Ekho Suverenizatsii', *Rossiiskaya Federatsiya Segodnya*, No. 3.

Börzel, T. (2000) 'From Competitive Regionalism to Co-operative Federalism: the Europeanization of the Spanish State of the Autonomies', *Publius: The Journal of Federalism*, 30(2): 17–42.

Brown, A. (1977) 'Introduction' in Brown, A. and Gray, J. (eds) (1977) *Political Culture and Political Change in Communist States*, London: MacMillan, pp. 1–25.

—— (1997) *The Gorbachev Factor*, Oxford: Oxford University Press.

—— (ed.) (2001a) *Contemporary Russian Politics: A Reader*, Oxford: Oxford University Press.

—— (2001b) 'Vladimir Putin and the Reaffirmation of Central State Power', *Post-Soviet Affairs*, 17(1): 45–55.

—— (2005) 'Conclusions', in Whitefield, S. (ed.) (2005) *Political Culture and Post-Communism*, London: Palgrave Macmillan, pp. 180–202.

Brown, A. and Gray, J. (eds) (1977) *Political Culture and Communist Studies*, London: Macmillan.

Brown, A. and Shevtsova, L. (eds) (2001) *Gorbachev, Yeltsin, and Putin: Political Leadership in Russia's Transition*, Washington, D.C.: Carnegie Endowment for International Peace.

Brubaker, R. (1996) *Nationalism Reframed. Nationhood and the National Question in the New Europe*, Cambridge UK: Cambridge University Press.

Bunin, I. (2002) 'Politicheskii Rezhim Vladimira Putina v Nachale Vtorogo Goda Pravleniya', Moscow Centre for Political Technology, October. Online. Available at www.cpt.ru/statint.php (accessed 20 April 2009).

—— (2005) 'V Poiskakh Novoi Sotsialnoi Roli Biznesa', Conference Materials, Asotsiatsiya Menedzherov Rossii, Moscow, 17 November, unpublished paper.

Bunin, I., Zudin, A., Makarenko, B. and Makarkin, A. (2002) 'Nachalo "Bolshoi Igry". Novy Izbiratelnyi Tsikl, ego Uchastniki, Intriga i Stsenarii', Moscow Centre for Political Technology, 19 July. Online. Available at www.politcom.ru/2002/p_pr6.php (accessed 20 April 2009).

Burgess, M. (1986) *Federalism and Federation in Western Europe*, New Hampshire: Croom Helm.

—— (2000) *Federalism and European Union: the Building of Europe, 1950–2000*, London: Routledge.

Burgess, M. and Gagnon, A.-G. (eds) (1993) *Comparative Federalism and Federation: Competing Traditions and Future Directions*, Toronto: University of Toronto Press.

Cardinal, L. (2004) 'The Limits of Bilingualism in Canada', *Nationalism and Ethnic Politics*, 10(1): 79–103.

Cassis, Y. (1997) *Big Business. The European Experience in the Twentieth Century*, Oxford: Oxford University Press.

Cespa, G. and Cestone, G. (2004) *Corporate Social Responsibility and Managerial Entrenchment*. Discussion Paper series, Paper No. 4648, London: Centre for Economic Policy Research.

Chapman, R. (1993) 'Structure, Process and the Federal Factor: Complexity and Entanglement in Federations', in Burgess, M. and Gagnon, A.-G. (eds) (1993) *Comparative Federalism and Federation: Competing Traditions and Future Directions*, Toronto: University of Toronto Press, pp. 69–94.

Chirikova, A. (2001) 'Rossiiskie Direktora i Regionalnaya Vlast. Poisk Optimalnykh Modelei Vzaimodeistviya', *Sotsiologicheskie Issledovaniya: SOCIS*, 11: 35–46.

—— (2003) 'Regionalnaya Vlast: Novye Protsessy i Novye Figury', in Chirikova, A. and Lapina, N. (eds) (2003) *Regionalnye Protsessy v Sovremennoi Rossii*, Moscow: INION RAN, pp. 89–115.

—— (2005) 'Regionalnaya Vlast v Rossii: Top Menedgery Kompanii kak Vlastnye Aktory', in Frukhtman, J. (ed.) (2005) *Regionalnaya Elita v Sovremennoi Rossii*, Moscow: Fond Liberalnaya Missiya, pp. 195–210.

Chirikova, A. and Lapina, N. (1999) *Regionalnye Elity v RF: Modeli Povedeniya i Politicheskie Orientatsii*, Moscow: INION RAN.

—— (2001) 'Regionalnaya Vlast i Reforma Rossiiskogo Federalisma: Stsenarii Politicheskogo Budushchego', *Sotsiologicheskie Issledovaniya: SOCIS*, 4: 16–27.

—— (2004) *Putinskie reformy i potentsial vliyaniya regionalnykh elit*, Moscow: Institut Kompleksnykh Sotsialnykh Issledovanii RAN.

Cohen, A. (1985) *The Symbolic Construction of Community*, London: Routledge.

—— (1974) *Two Dimensional Man: an Essay on the Anthropology of Power and Symbolism in Complex Society*, London: Routledge & Kegan Paul.

Colton, T.J. and McFaul, M. (2002) 'Are Russians Undemocratic?', *Post-Soviet Affairs*, 18(2): 91–121.

Colton, T., Goldman, M., Saivetz, C., and Szporluk, R. (2005) 'Russia in the Year 2004', *Post-Soviet Affairs*, 21(1): 1–25.

Conroy, M.S. (1998) *Emerging Democracy in Late Imperial Russia. Case Studies on Local Self-government (The Zemstvos), State Duma Elections, the Tsarist Government, and the State Council before and During World War I*, Colorado: University of Colorado Press.

Corbridge, S. (1995) 'Federalism, Hindu, Nationalism and Mythologies of Governance in Modern India', in Smith, G. (ed.) (1995) *Federalism: The Multiethnic Challenge*, London: Longman, pp. 101–27.

Corwin, J.A. (2000) 'Vulnerability of Governors to Dismissal Questioned', *RFE/RL Russian Federation Report*, 2: 21 (7 June).

Covell, M. (1987) 'Federalization and Federalism', in Bakvis, H. and Chandler, W. (eds) (1987) *Federalism and the Role of the State*, Toronto: University of Toronto Press, pp. 82–157.

Cox, G. (1987) *The Efficient Secret*, New York: Cambridge University Press.

Dahl, R. (1956) *A Preface to Democratic Theory*, Chicago: University of Chicago Press.

—— (1986) *Democracy, Liberty and Equality*, Oslo: Norwegian University Press.

—— (1989) *Democracy and its Critics*, London: Yale University Press.

Davis, R. (1978) *The Federal Principle: A Journey Through Time in Quest of Meaning*, Berkley: University of California Press.

Dicey, A.V. (1967) *Introduction to the Study of the Law of the Constitution*, 10th edn, London: Macmillan Press.

Dimock, E. (1949) *Business and Government*, New York: Henry Holt.

Drobizheva, L. (ed.) (1998) *Asymmetrichnaya Federatsiya, Vzglyad is Tsentra, Respublik i Oblastei*, Moscow: Institute of Ethnology and Anthropology RAN.

Drozdov, Yu. and Fartsyshev, V. (2001) *Yurii Andropov i Vladimir Putin Na Puti k Vozrozhdeniyu*, Moscow: OLIMPA-PRESS.

Duchacek, I.D. (1970) *Comparative Federalism*, New York: Holt, Rinehart and Winston.

—— (1986) *Comparative Federalism: The Territorial Dimension of Politics*, Boulder, CO: Westview Press.

Duverger, M. (1964) *Political Parties: Their Organisation and Activity in the Modern State*, 3rd edn, London: Methuen.

Dyshekova, M. (2003) 'Konfederatsiya po-Shveitsarski', *Rossiiskaya Federatsiya Segodnya*, 2(1).

Eckstein, H. (1998) 'Congruence Theory Explained', in Eckstein, H., Fleron F.J. Jr., Hoffman, E.P., and Reisinger, W.M. (eds) (1998) *Can Democracy Take Root in Post-Soviet Russia? Explorations in State-Society*, Maryland: Rowman & Littlefield, pp. 3–34.

Elaigwu, J.I. (2002) 'Federalism in Nigeria's New Democratic Polity', *Publius: the Journal of Federalism*, 32(2): 73–95.

Elazar, D.J. (1987) *Exploring Federalism*, Tuscaloosa: University of Alabama Press.

—— (1997) 'Contrasting Unitary and Federal Systems', *International Political Science Review*, 18(3): 237–51.

Elliot, J. (1997) *Transfer Pricing: Lessons form Australia 'Striking a Balance'*, Southampton: School of Management University of Southampton.

Emizet, K.N. and Hesli V.L. (1995) 'The Disposition to Secede: An Analysis of the Soviet Case', *Comparative Political Studies*, 27(4): 493–536.

Fairbrass, J. (2006) Sustainable Development, Corporate Social Responsibility and Eureopanisation of the UK Business Actors: Preliminary Findings. Working Paper No. 06/02, February, University of Bradford: Working Paper series.

Farmer, R.N. and Hogue, D. (1973) *Corporate Social Responsibility*, USA: Science Research Associates.

Filipov, M., Ordeshook, P., and Shvetsova, O. (2004) *Designing Federalism: a Theory of Self-Sustainable Federal Institutions*, Cambridge: Cambridge University Press.

Fleiner, T. (2002) 'Recent Developments of Swiss Federalism', *Publius: Journal of Federalism*, 32(2): 97–124.

FOM (2005) '*Ugorza Raspada Rossii: Realnaya ili Mnimaya?*' 21 April, http://bd.fom. ru/report/map/d051627 (accessed 2 November 2006).

Franck, T.M. (1968) *Why Federations Fail. An Inquiry into the Requisites of Successful Federalism*, New York: New York University Press.

Friedman, M. (1970) 'The Social Responsibility of Business is to Increase its Profits', *New York Times Magazine*, 13 September.

Friedrich, C.J. (1968) *Trends of Federalism in Theory and Practice*, London: Pall Mall Press.

Frukhtman, J. (ed.) (2005) *Regionalnaya Elita v Sovremennoi Rossii*, Moscow: Fond Liberalnaya Missiya.

Gaidar, Y. (2007) 'My Sidim Na Porokhovoi Bochke', *Novoe Vremya*, 22 February.

Galligan, B. (1995) *A Federal Republic*, Cambridge: Cambridge University Press.

Gaman-Golutvina, O. (2004) 'Rossiiskie Elity: Vzglyad iz Regionov', *Politiya*, 26 February.

Gazizova, E. (2007a) 'Zhizn' s Protyanutoi Rukoi', *Ekspert*, 19 February.

—— (2007b) 'Osobye Otnosheniya', *Ekspert*, 5 June.

Geertz, C. (1975) *The Interpretation of Cultures*, New York: Basic Books.

Gelman, V. (2000) 'The Politics of Russia's Regions: a Comparative Perspective', in Harter, S. and Easter, G. (eds) (2000) *Shaping the Economic Space in Russia: Decision making Processes, Institutions and Adjustement to Change in the El'tsin Era*, Aldershot: Ashgate, pp. 227–48.

—— (2001) 'The Future of Regional Electoral Reform', *EWI Russian Regional Report* 6, No. 33, September 26. Online. Available . (accessed 20 April 2009).

—— (2002) 'The Rise and Fall of Federal Reform in Russia', *Program on New Approaches to Russian Security Policy Memo Series*, No. 238, January 25. Online. Available at www.csis.org/media/csis/pubs/pm_0238.pdf (accessed 20 April 2009).

Gelman, V., Ryzhenkov, S., and Bri, M. (2000) *Rossiya Regionov: Transformatsiya Politicheskikh Rezhimov*, Moscow: Ves Mir.

Gibson, J. (2001) 'The Russian Dance with Democracy', *Post-Soviet Affairs*, 17(2): 101–28.

Gilliomee, H. (2004) The Rise and Possible Demise of Afrikaans and Public Language, *Nationalism and Ethnic Politics*, 10(1): 25–58.

Golosov, G. (2003) *Political Parties in the Regions of Russia: Democracy Unclaimed*, Boulder, CO: Lynne Reinner.

Goode, J.P. (2004) 'The Push for Regional Enlargement in Putin's Russia', *Post-Soviet Affairs*, 20(3): 219–57.

Gorenburg, D. (1999) 'Regional Separatism in Russia: Ethnic Mobilisation or Power Grab?', *Europe-Asia Studies*, 51(2): 245–74.

—— (2001) 'Nationalism for the Masses: Popular Support for Nationalism in Russia's Ethnic Republics', *Europe-Asia Studies* 53(1): 73–104.

Gray, J. (1979) 'Conclusions', in Brown, A. and Gray, J. (eds) (1979) *Political Culture and Political Change in Communist States*, London: Macmillan, pp. 253–73.

Greenstein, F. (1969) *Personality and Politics*, Chicago: Markham Publishing Company.

Griffiths, A.L. (2002) *Handbook of Federal Countries 2002*, Ithaca: McGill University Press.

Gromov, A. (2007) 'Skandal na Rovnom Meste', *Expert*, 19 February.

Guibernau, M. (1996) *Nationalisms. The Nation-State and Nationalism in the Twentieth Century*, Cambridge, UK: Polity Press.

Guseinov, V. (1999) *Ot El'tsina k…? Pyanyashchii Durman Vlasti*, Moscow: OLMA-PRESS.

Hahn, G. (2003) 'The Impact of Putin's Federative Reforms on Democratisation in Russia', *Post-Soviet Affairs*, 19(2): 114–53.

Hahn, J. (2004) 'St. Petersburg and the Decline of Local Self-Government in Post-Soviet Russia', *Post-Soviet Affairs*, 20(2): 107–32.

Hale, H.E. (2000) 'The Parade of Sovereignties: Testing Theories of Secession in the Soviet Setting', *The British Journal of Political Science*, 30(1): 31–56.

—— (2003) 'Explaining Machine Politics in Russia's Regions: Economy, Ethnicity, and Legacy', *Post-Soviet Affairs*, 19(3): 228–63.

—— (2004) 'Divided We Stand. Institutional Sources of Ethnofederal State Survival and Collapse', *World Politics*, 56: 165–93.

Hale, H., McFaul, M., and Colton, T. (2004) 'Putin and the "Delegative Democracy" Trap: Evidence from Russia's 2003–4 Elections', *Post-Soviet Affairs*, 20(4): 285–319.

Hanson, P. and Teague, E. (2005) 'Big Business and the State in Russia', *Europe-Asia Studies*, 57(5): 657–80.

Hart, S. (1995) 'A Natural Resource-Based View of the Firm', *Academy of Managment Review*, 20: 986–1014.

Hesli, V.L. and Reisinger, W.M. (eds) (2003) *The 1999–2000 Elections in Russia: Their Impact and Legacy*, New York, NY: Cambridge University Press.

Hicks, U. (1978) *Federalism: Failure and Success*, London: Macmillan Press.

Horowitz, D. (1985) *Ethnic Groups in Conflict*, Los Angeles: Univesity of California Press.

Hyde, M. (2001) 'Putin's Federal Reforms and their Implications for Presidential Power in Russia', *Europe-Asia Studies*. 53 (5): 719–43.

Ivanov, V.N. and Yarovoi, O.Y. (2000) *Rossiiskii Federalism: Stanovleniye i Razvitiye*, Moscow: Institute of Socio-Political Research (RAN).

Jeffery C. and Hough, D. (2003) 'Regional Elections in Multi-Level Systems', *European Urban and Regional Studies*, 10 (3): 199–212.

Jones, M. (1980) 'Corporate Social Responsibility Revisited, Redefined', *California Management Review*, 22(2): 59–67.

Kahn, J. (2000) 'The Parade of Sovereignties: Establishing the Vocabulary of the New Russian Federalism', *Post-Soviet Affairs*, 16(1): 58–89.

—— (2001) 'What is the New Russian Federalism?', in Brown, A. (ed.) (2001) *Contemporary Russian Politics: A Reader*, Oxford: Oxford University Press, pp. 374–83.

—— (2002) *Federalism, Democratization and the Rule of Law in Russia*, Oxford: Oxford University Press.

Kazakov, M.A. (2001) 'Regionalnye Elity: Sravnitelnye Aspekty Evolutsii', Nizhnii Novgorod Research Foundation, 20 December, unpublished paper.

Khakimov, R. (1997) 'Asymmetrichnost Rossiiskoi Federatsii: Vzglyad iz Tatarstana', in Zakharov, A. (ed.) (1997) *Asymmetrichnost Federatsii*, Moscow: TACIS, pp. 61–76.

—— (1998) 'Ob Osnovakh Asymmetrichnosti Rossiiskoi Federatsii', in Drobizheva, L. (ed.) (1998) *Asymmetrichnaya Federatsiya, Vzglyad is Tsentra, Respublik i Oblastei*, Moscow: Institut Sotsiologii RAN, pp. 37–48.

Kharkhordin, O. (2005) *Main Concepts of Russian Politics*, New York: University Press of America.

Khristenko, V. (2002) 'Razvitie byudzhetnogo federalizma v Rossii: Itogi 1990-kh godov i zadachi na perspektivu', *Voprosy Ekonomiki*, 2 (February): 4–18.

Kichedzhi, V. (2003) 'Nedaleko ot Moskvy', *Buziness Obozrenie*, 3: 16–19.

Kimura, M. (2003) 'Memories of Massacre: Violence and Collective Identity in the Narratives of the Nellie Incident', *Asian Ethnicity*, 4(2): 225–39.

King, P. (1982) *Federalism and Federation*, London & Canberra: Croom Helm.

Klimanov, V. (2001) 'Regionalnoe Sotrudnichestvo', in Petrov, N. (ed.) (2001) *Regiony Rossii v 1999 godu: Ezhegodnoe Prilozhenie k 'Politicheskomu Almanakhu Rossii'*, Moscow: Gendalf, pp. 124–9.

Klimentev, E. (2004) 'V Poiskakh Pravovoi Zashchity Kulturno-Yazykovykh Interesov Karel, Vepsov, Finnov Respubliki Kareliya', *Kazanskii Federalist*, 3(11): 70–108.

Kovalskaya, G. (2001), 'Posredstvom Nakhrapa. Nazdratenko kak Simvol Vremeni', *Itogi*, No. 5.

Kozlov, A.E. (1996) *Federativnye Nachala Gosudarstvennoi Vlasti v Rossii*, Moscow: INION RAN.

Kozlov, M. (1998) 'Novye Tendentsii v Otnosheniyakh Subektov Federatsii s Organami Vlasti Tsentra', in Drobizheva, L. (ed.) (1998) *Asymmetrichnaya Federatsiya. Vzglyad iz Tsentra, Respublik i Oblastei*, Moscow: Institut Sotsiologii RAN, pp. 49–57.

Kramer, L. (1994) 'Understanding Federalism', *Vanderbilt Law Review*, 47(1): 1485–561.

Kravtsov, D. (2003) 'Troe v Lodke, ne Schitaya Prizrakov', *Ekspert*, 23 June.

Kryshtanovskaya, O. (2002) 'Rezhim Putina: Liberalnaya Militokratiya?', *Pro et Contra*, 7(4): 158–80.

—— (2003) 'Politicheskie Reformy Putin i Elita', *Ekonomika i Obshchestvo*, 4–5: 3–50.
Kryshtanovskaya, O. and White, S. (2003) 'Putin's Militocracy', *Post-Soviet Affairs*, 19(4): 289–306.
Kuzmin, A. (2001) 'Regionalnye Elity Podelyat "Sinyaki i Shishki?"', *Ekonomika Rossii XXI Vek*, No. 4 (4), October.
Kynev, A. (2004) 'Nachalo Poslednei Volny?', *Politicheskii Zhurnal*, No. 36 (39)/05, October.
—— (2005) 'Vybory Regionalnykh Zakonodatelnykh Sobranii Oseni 2004 Goda: Izbiratel Ishchet Novykh Geroev', Fund Indem Publications, Autumn. Online. Available at www.indem.ru/idd2000/anal/Kinev/OsenZS05.htm (accessed 20 April 2009).
—— (2008) 'Osnovnye Tendentsii Regionalnoi Politicheskoi Zhizni Letom 2007 Goda', *Carnegie Centre*. Online. Available at http://monitoring.carnegie.ru/2008/01/analytics/aleksandr-kynev-osnovnye-tendencii-regionalnoj-politicheskoj-zhizni-v-oktyabre-2007/ (accessed 30 September 2008).
Lallemand, J.-C. (1999) 'Politics for the Few: Elites in Bryansk and Smolensk', *Post-Soviet Affairs*, 15(4): 312–35.
Lane, D. and Ross, C. (1999) *The Transformation from Communism to Capitalism: Ruling Elites from Gorbachev to Yeltsin*, New York: St. Martin's Press.
Lankina, T. (2004) *Governing the Locals. Local Self-Government and Ethnic Mobilisation in Russia*, Oxford: Rowman & Littlefield.
Lapina, N. (2005) 'Biznes i Vlast: Sokhranitsya li Regionalnoe Mnogoobrazie?', in Frukhtman, J (ed.) (2005) *Regionalnaya Elita v Sovremennoi Rossii*, Moscow: Fond Liberalnaya Missiya, pp. 65–77.
Leontev, M. (2004) 'Soyuz Mecha i Orala: Novaya Obedinennaya Oppozitsiya za Rossiyu Bez Putina', *Izvestiya*, 25 February.
Levada, Yu. (2000) 'Ispytatelnyi Srok. Vremya Poshlo. Nastroeniya i Mneniya. Iyun 2000', *Nezavisimaya Gazeta*, 12 July.
Levitt, T. (1958) 'The Dangers of Social Responsibility', *Harvard Business Review*, 36(5): 41–50.
Lijphart, A. (1984) *Democracies: Patterns of Majoritarian and Consensus Government in Twenty-One Countires*, New Haven: Yale University Press.
—— (1999) *Patterns of Democracy: Government Forms and Performance in Thirty-Six Countries*, New Haven, Yale University Press.
Lipset, S.M. (1960) *Political Man*, New York: Doubleday.
Litovchenko, S. (2004) *Doklad o Sotsialnykh Investitsiakh v Rossii za 2004 God*, Moscow: Assotsiatsiya Menedgerov Rossii. Unpublished paper.
Livingston, W. (1956) *Federalism and Constitutional Change*, Oxford: Oxford University Press.
Loginovskii, S. (1997) 'K Novomu Territorialnomu Ustroistvu Rossii', *Polis*, 5: 140–6.
Lowi, T.J., and Ginsberg, B. (2000) *American Government: Freedom and Power*, 6th edn, New York: W.W. Norton.
Lukin, A. (2000) *The Political Culture of the Russian 'Democrats'*, Oxford: Oxford University Press.
—— (2001) 'Putin's Regime: Restoration or Revolution?', *Problems of Post-Communism*, 48(4): 38–48.
Lysenko, V. (1995) *Razvitie Federativnykh Otnoshenii v Sovremennoi Rossii*, Moscow: Institut Sovremennoi Politiki.
—— (1997a) 'Razvitie Federatsii i Konstitutsiya Rossii', *Gosudarstvo i Pravo*, 8: 14–20.
—— (1997b) 'Razdelenie Vlastei i Opyt Rossiiskoi Federatsii' in Zakharov, A. (ed.) (1997) *Asimmetrichnost Federatsii*, Moscow: TACIS, pp. 13–35.
—— (1998) 'Dogovornye Otnosheniya kak Faktor Obostreniya Protivorechii Kraev i Oblastei s Respublikami', in Drobizheva, L. (ed.) (1998) *Asymmetrichnaya federat-*

siya: Vzglyad iz tsentra, respublik i oblastei, Moscow: Institute of Ethnology and Anthropology RAN, pp. 32–55.

—— (2002) 'Razvitie Federalnykh Okrugov i Budushchee Federativnogo Ustroistva Rossii', *Kazanskii Federalist*, 2: 38–50.

—— (2005) 'Naznachenie Gubernatorov – Put k Odnopartiinoi Sisteme', *Nezavisimaya Gazeta*, 24 March.

Makarkin, A. (2002) 'Kak Semya Delaet Biznes: Roman Abramovich i Oleg Deripaska v Interere Svoego Dela', Moscow Centre for Political Technology, 28 April. Online. Available at www.cpt.ru/statint.php (accessed 20 April 2009).

—— (2004) 'Distsiplinirovannaya Duma', Moscow Centre for Political Technology, 19 January. Online. Available at www.politcom.ru/2004/pvz339.php (accessed 20 April 2009).

—— (2007) 'Spravedlivaya Rossiya: Serebro ili Bronza', *Ezhednevnyi Zhurnal*, 1 August.

McWilliams, A., Siegel, D.S., and Wright, P.M. (2006) 'Corporate Social Responsibility: Strategic Implications', *Journal of Management Studies*, 43(1): 1–19.

Mendras, M. (1999) 'How Regional Elites Preserve their Power', *Post-Soviet Affairs*, 15(4): 295–311.

Mikhailov, A.P. (2001) 'Modelirovanie Rossiiskoi Vlasti', *Sotsiologicheskie Issledovaniya SotsIS*, 5: 12–20.

Mikheev, S. (2001) 'Reorganizatsiya MVD Bet po Regionalnym Lideram', Publications of the Moscow Centre for Political Technology, 29 September.

—— (2002) 'Tsentr-Natsionalnye Respubliki: Otnosheniya na Novom Etape', Moscow Centre for Political Technology, 30 October. Online. Available at www.politcom.ru/2002/p_region31.php (accessed 20 March 2008).

Moscovici, S. (1977) *Social Influence and Social Change*, US: Academic Press.

Nemytykh, Y. (2004) 'Administrativnaya Petlya', *Expert*, 28 June.

Nicholson, M. (1999) *Towards a Russia of the Regions*, Oxford: Oxford University Press.

Niou, S. and Ordeshook, P. (1998) 'Alliances versus Federations: An Extension of Riker's Analysis of Federal Formation', *Constitutional Political Economy*, 9(4): 271–88.

Olshanskii, D. (2000) 'Dezintegratsiya: Novye Simptomy Staroi Bolezni', *Pro et Contra*, 5(1): 36–42.

Ordeshook, P. (1996) 'Russia's Party System: Is Russian Federalism Viable?', *Post-Soviet Affairs* 12(3): 195–217.

Ordeshook, P. and Shvetsova, O. (1997) 'Federalism and Constitutional Design', *Journal of Democracy*, 8(1): 27–42.

Orlov, D. (2002) 'Oligarkhi Vozvrashchayutsya', *Vremia-MN*, 8 October.

Ortino, S., Zagar, M., and Mastny, V. (2005) *The Changing Faces of Federalism. Institutional Reconfiguration in Europe from East to West*, Manchester: Manchester University Press.

Orttung, R. (2004) 'Key Issues in the Evolution of the Federal Okrugs', in Reddaway, P. and Orttung, R. (eds) (2004) *The Dynamics of Russian Politics: Putin's Reform of Federal-Regional Relations, Volume 1*, Oxford: Rowman & Littlefield, pp. 19–53.

Orttung, R. and Reddaway, P. (2004) *The Dynamics of Russian Politics: Putin's Reform of Federal-Regional Relations*, Oxford: Rowman & Littlefield.

Ostrom, V. (1991) *The Meaning of American Federalism: Constituting a Self-Governing*, San Francisco: Institute for Contemporary Studies.

Ostrow, J.M. (2002) 'Conflict Management in Russia's Federal Institutions', *Post-Soviet Affairs*, 18(1): 49–70.

Pappe, Y. (2005) 'Otnosheniya Federalnoi Ekonomicheskoi Elity i Vlasti v Rossii v 2000–2004 godakh: Tormozhenie v Tsentre i Novaya Strategiya v Regionakh', in Frukhtman, J. (ed.) (2005) *Regionalnaya Elita*, Moscow: Liberal Mission Foundation Press, pp. 77–89.

Pavlov, A. and Perrie, M. (2003) *Ivan the Terrible. Profiles in Power*, London: Longman Press.
Pechenev, V. (2001) *Vladimir Putin: Poslednii Shans Rossii?*, Moscow: INFRA-M.
Peregudov, P. (2003) *Korporatsii, Obshchestvo, Gosudarstvo*, Moskva: Nauka.
Petro, N. (1995) *The Rebirth of Russian Democracy*, Cambridge Massachusetts: Harvard University Press.
—— (2004) *Crafting Democracy. How Novgorod Has Coped with Rapid Social Change*, USA: Cornell University Press.
Petrov, N. (ed.) (1999) *Regiony Rossii v 1998 godu: Ezhegodnoye Prilozheniye k 'Politicheskomu Almanakhu Rossii'*, Moscow: Carnegie Foundation, Gendalf.
—— (2000a) Federalizm po-Rossiiski, *Pro et Contra*, 5(1): 7–33.
—— (2000b) 'Broken Pendulum: Recentralization under Putin', *Program on New Approaches to Russian Security Policy Memo Series*, No. 159, November. Online. Available at www.csis.org/ruseura/ponar (accessed 20 April 2009).
—— (ed.) (2000c) *Regiony Rossii v 1998 godu: Ezhegodnoe Prilozhenie k 'Politicheskomu Almanakhu Rossii'*, Moscow: Carnegie Foundation, Gendalf.
—— (ed.) (2001) *Regiony Rossii v 1999 godu: Ezhegodnoe Prilozhenie k 'Politicheskomu Almanakhu Rossii'*, Moscow: Carnegie Foundation, Gendalf.
—— (2006) 'Polpredy v Sisteme Federalnoi Ispolnitelnoi Vlasti', in Khakimov, R. (ed.) (2006) *Politiko-pravovye Resursy Federalizma v Rossii*, Kazan, Kazanskii Institut Federalizma, pp. 75–107.
Petrov, N. and Titkov, A. (1999) 'Politicheskaya Zhizn v Regionakh', in Petrov, N. (ed.) (2000) *Regiony Rossii v 1998 godu: Ezhegodnoe Prilozhenie k 'Politicheskomu Almanakhu Rossii'*, Moscow: Gendalf.
Pipes, R. (1974) *Russia under the Old Regime*, New York: Scribner.
Plasschaert, S.R.F. (1979) *Transfer Pricing and Multinational Corporations: An Overview of Concepts, Mechanisms and Regulations*, England: European Centre for Study and Information on Multinational Corporations (ECSIM).
Plyais, Ya. (1998) *Rossiiskaya gosudarstvennost: Opyt, sovremennoe sostoyanie, perspektivy reformirovaniya*, Moscow: Finansovaya Akademiya pri Prezidente Rossiiskoi Federatsii.
Polin, D. (2005) 'Ochered na Ukrupnenie', *Ekspert*, 3 October.
Poltavchenko, G. (2003) Internet Conference with Poltavchenko, G. S., plenipotentiary representative in the Central Federal District, 19 April. Online. Available at www.garweb.ru/conf/polpred_cfo/20030429/index.htm (accecssed 20 April 2009).
Powell, B. (1982) *Contemporary Democracies: Participation, Stability, Violence*, Cambridge Mass: Harvard University Press.
Przeworski, A., Alvarez, M., Cheibub, J., and Limongi, F. (2000) *Democracy and Development: Political Institutions and Well-Being in the World, 1950–1990*, New York: Cambridge University Press.
Putnam, R. (1973) *The Beliefs of Politicians*, New Haven and London: Yale University Press.
—— (1993) *Making Democracy Work: Civic Traditions in Modern Italy*, Princeton: Princeton University Press.
Pye, L.W. and Verba, S. (1965) *Political Culture and Political Development*, Princeton, NJ: Princeton University Press.
Radygin, A. (2004) 'Uklonenie ot Neuplaty Nalogov', Fond Liberalnaya Missiya, 10 March. Online. Available at www.liberal.ru/libcom.asp?Num=57 (accessed on 24 September 2007).
Rath, S. (1984) *Federalism Today: Approaches, Issues, and Trends*, New Delhi: Sterling Publisher.
Remington, T. (2003) 'Majorities without Mandates: The Russian Federal Council since 2000', *Europe-Asia Studies*, 55(5): 667–91.
Riker, W.H. (1964) *Federalism: Origins, Operation, Significance*, Boston: Little Brown.

—— (1975) 'Federalism', in Greenstein, F. and Polsby, N.W. (eds) (1975) *Handbook of Political Science 5: Governmental Institutions and Processes*, Reading, Mass: Addison Wesley, pp. 93–172.

Robbins, R. Jr. (1987) *The Tsar's Viceroys. Russian Provincial Governors in the Last Years of the Empire*, Cornell University Press: Ithaca and London.

Rokeach, M. (1960) *The Open and Closed Mind*, New York: Basic Books.

Rokkan, S. (1970) *Citizens, Elections, Parties*, New York: McKay.

Rokkan, S. and Urwin, D. (1982) *The Politics of Territorial Identity*, London: Sage.

—— (1983) *Economy, Territory, Identity*, London: Sage Publications.

Roland, G. (2006) 'The Russian Economy in the Year 2005', *Post-Soviet Affairs*, 22(1): 90–8.

ROMIR (2001) 'Elita ob Osnovnykh Ugrozakh Bezopasnosti Rossii', 4 October. Online. Available at www.4p.ru/index.php?page=930&tmpl=print (accessed 2 November 2006).

—— (2003) 'Chto Volnuet Rossiiskikh Grazhdan?', 1 December. Online. Available at http://rmh.ru/news/res_results/84.html (accessed 2 November 2006).

Rose, R. and Munro, N. (2002) *Elections Without Order*, New York, NY: Cambridge University Press.

Ross, C. (2002) *Federalism and Democratisation in Russia*, Manchester: Manchester University Press.

—— (2003) 'Putin's Federal Reforms and the Consolidation of Federalism in Russia: One Step Forward, Two Steps Back', *Communist and Post-Communist Studies*, 36: 29–47.

—— (ed.) (2004) *Russian Politics Under Putin*, Manchester: Manchester University Press.

Rubchenko, M. and Shokhina, E. (2003) 'Gryaznaya Troika', *Ekspert*, 1 December.

Rugman, A. and Eden, L. (1985) *Multinational and Transfer Prices*, Sidney: Croom Helm Australia.

Rumyantsev, O. (1995) 'K rossiiskomu soyuzu respublik i zemel', *Obozrevatel*, 10 (69): 2–8.

Ryabov, A. (2007) 'Osnovnye tendentsii rossiiskogo izbiratelnogo tsikla: 2007–2008', Publications of the Moscow Carnegie Centre, 26 April. Online. Available at www.carnegie.ru/ru/pubs/media/76117.htm (accessed 20 December 2007).

Safran, W. (1999) 'Politics and Language in Contemporary France: Facing Supra-national and Infra-National Challenges', *International Journal of the Sociology of Language*, 137: 39–66.

—— (2004) 'The Political Aspects of Language', *Nationalism and Ethnic Politics*, 10(1): 1–14.

Sakwa, R. (2002) 'Federalism, Sovereignty and Democracy', in Ross, C. (ed.) (2002) *Regional Politics in Russia*, Manchester: Manchester University Press, pp. 1–22.

—— (2003) 'Elections and National Integration in Russia', in Helsi, V. and Reisinger, W. (eds) (2003) *The 1999–2000 Elections in Russia*, Cambridge: Cambridge University Press, pp. 121–42.

—— (2004) *Putin: Russia's Choice*, London: Routledge.

—— (2005) 'Partial Adaptation and Political Culture', in Whitefield, S. (ed.) (2005) *Political Culture and Post-Communism*, London: Palgrave Macmillan, pp. 42–64.

Samuels, D. (2000) 'Federalism and Democratic Transitions: the "New" Politics of the Governors in Brazil', *Publius: The Journal of Federalism*, 30(2): 43–61.

Sawer, G. (1976) *Modern Federalism*, London: Pitman Publishing.

—— (1977) 'New Federalism', in Dean, J. (ed.) (1977) *The Politics of New Federalism*, Sidney: Australian Political Science Association, Bridge Printery, pp. 15–20.

Sedlak, A. (2007) 'Delo Dlya Kandidata', *Ekspert*, 28 November 2007.

Shakhrai, S. (1994) 'O Kontseptsii Gosudarstvennoi Regionalnoi Politiki', *Nezavisimaya Gazeta*, May 12.

Shevtsova, L. (2004) 'Vpered v Proshloe ili Manifest Stagnatsii', *Izvestiya*, 25 February.

Shleifer, A. and Treisman, D. (2000) *Without a Map. Political Tactics and Economic Reform in Russia*, Cambridge, Mass: MIT Press.

Shmarov, A. and Stolyarov, B. (2007) 'Public-Private Partnership: Perevod na Russkii', *Ekspert*, 25 December.

Shugart, M. and Carey, J. (1992) *Presidents and Assemblies*, Cambridge: Cambridge University Press.

Shugart M. and Mainwaring, S. (1997) *Presidentialism and Democracy in Latin America*, Cambridge: Cambridge University Press.

Simeon, R. (ed.) (1985) *Intergovernmental Relations*, Toronto: University of Toronto Press.

Skinner, Q. (1978) *The Foundations of Modern Political thought. Volume Two: The Age of Reformation*, Cambridge: Cambridge University Press.

—— (1989) 'The State', in Ball, T., Farr, J., and Hanson R. (eds) (1989) *Political Innovation and Conceptual Change*, Cambridge, UK: Cambridge University Press.

—— (2002) *Visions of Politics: Regarding Method. Vol. 1*, Cambridge, UK: Cambridge University Press.

Smiley, D. (1972) *Canada in Question: Federalism in the Seventies*, Canada: McGraw Hill.

Smirnyagin, L. (2001) 'Federalizm po Putinu ili Putin po Federalizmu? (Zheleznoi Pyatoi)', *Briefing*, 3(3): 1–4.

Smith, A. (1976) 'Introduction: the Formation of Nationalist Movements', in Smith, A. (ed.) (1976) *Nationalist Movements*, New York: Macmillan Press, pp. 1–30.

—— (1995) *Nations and Nationalism in a Global Era*, Cambridge, UK: Polity Press.

Smith, G. (1998) 'Russia, Multiculturalism and Federal Justice', *Europe-Asia Studies*, 50(8): 1393–411.

—— (ed.) (1995) *Federalism: the Multi-ethnic Ethnic Challenge*, London: Longman.

Solnick, S. (1995) 'Federal Bargaining in Russia', *East European Constitutional Review*, 4(4): 52–6.

—— (1996) 'The Political Economy of Russian Federalism. A Framework for Analysis', *Problems of Post-Communism*, 43(6): 13–25.

—— (1998) 'Gubernatorial Elections in Russia, 1996–1997', *Post-Soviet Affairs*, 14(1): 48–80.

—— (2000) 'Putin and the Provinces', *Program on New Approaches to Russian Security Policy* PONARS *Memo Series*, No. 115 (April). Online. Available at www.csis. org/component/option,com_csis_pubs/task,view/id,2407/type,0/ (accessed 20 April 2009)

Spiro, H. (1959) *Government by Constitution*, New York: Random House.

Stanovaya, T. (2005) 'Tsentr i regiony: God posle otmeny pryamykh gubernatorskikh vyborov', *politcom.ru*, 29 September. Online. Available . (accessed 30 September 2008).

Stepan, A. (1999) 'Federalism and Democracy: Beyond the U.S. Model', *Journal of Democracy*, 10(4): 19–34.

—— (2000) 'Russian Federalism in Comparative Perspective', *Post-Soviet Affairs*, 16(2): 133–76.

Stoner-Weiss, K. (1997) *Local Heroes: The Political Economy of Russian Regional Governance*, Princeton: Princeton University Press.

—— (2002) 'Central Governing Incapacity and the Weakness of Political Parties: Russian Democracy in Disarray', *Publius: The Journal of Federalism*, 32(2): 125–46.

Suberu, R. (2001) Federalism and Ethnic Conflict in Nigeria, Washington DC: United States Institute of Peace Press.

Swenden, W. (2002) Asymmetric Federalism and Coalition-Making in Belgium, *Publius: the Journal of Federalism*, 32(3): 67–87.

Swidler, A. (1998) 'Culture and Social Action', in Smith, P. (ed.) (1998) *The New American Cultural Sociology*, Cambridge: Cambridge University Press.

Tang, R. (1979) *Transfer Pricing in the United States and Japan*, New York: Praeger Publishers.

Timofeyev, I. (2004) 'The Development of Russian Liberal Thought since 1985', in Brown, A. (ed.) (2004) *The Demise of Marxism-Leninism in Russia*, Basingstoke: Palgrave Macmilllan, pp. 51–119.

Tomberg, I. (2002) 'Regionalnye Lobby Nakanune Vtorogo Chteniya', Moscow Centre for Political Technology, 15 November 2002, available at www.cpt.ru/statint.php.

Tompson, W. (2005a) 'Putting *Yukos* in Perspective', *Post-Soviet Affairs*, 21(2): 159–81.

—— (2005b) 'The Political Implications of Russia's Resource-Based Economy', *Post-Soviet Affairs*, 21(4): 335–59.

Toshchenko, Zh. (1994) 'Asymmetriya kak Printsip Natsionalnogo i Federativnogo Stroitelstva' in Zh. Toshchenko, *Postsovetskoe Prostranstvo: Suverenizatsiya i Integratsiya*, Moscow: Etnosotsiologichekie Ocherki, RGGU, pp. 182–91.

Treisman, D. (1997) 'Russia's "Ethnic Revival". The Separatist Activism of Regional Leaders in a Postcommunist Order', *World Politics*, 49 (January), pp. 212–49.

Turovsky, R. (2001) 'V Idealnoi Federatsii Polpredy ne Nuzhny', *Rossiya*, 19 June.

—— (2002) 'Gubernatory i Oligarkhi: Istoriya Otnoshenii', Publications of the Moscow Centre for Political Technology, 26 February. Online. Available at www.politcom.ru/2002/aaa_c_vl3.php (accessed 20 April 2009).

—— (2004) 'Novaya i Khorosho Zabytaya Staraya Elektoralnaya Geografiya', Moscow Centre for Political Technology, January 14. Online. Available at www.politcom.ru/2004/obsh_vibor11.php (accessed 2 November 2006).

—— (2005) 'Putinskaya Pyatiletka v Regionalnoi Politike ili Beg po Krugu', Institute for Regional Policy, 28 December. Online. Available at http://regionalistica.ru/library/rft37.php (last accessed 20 December 2007).

—— (2007a) 'Regionalnye Ekonomiki i Regionalnyi Separatizm', Moscow Centre for Political Technology, 23 October 2007. Online. Available at www.politcom.ru/article.php?id=5246 (accessed 5 February 2008).

—— (2007b) 'The Mechanism of Representation of Regional Interests at the Federal Level in Russia: Problems and Solutions', *Perspectives on European Politics and Society*, 8(1): 73–97.

Turovsky, R. and Titkov, A. (2005) 'Liberalnyi Populizm Zavoevyvaet Krasnye Regiony', *gazeta.ru*. March. 30. Online. Available at www.gazeta.ru/comments/2005/03/30_x_261025.shtml (accessed 2 November 2006).

Vasileva, T. (2001) Analytical Report of the Institute of Law and Public Policy, 5(10): 1–18.

Verkhoturov, D. (2007) 'Nekonkretnoe Obedinenie', *Ekspert*, 22 January.

Verney, D. (1959) *The Analysis of Political Systems*, London: Compton Printing Works.

Vile, M.J.C. (1977) 'Federal Theory and the "New Federalism"', in Dean, J. (ed.) (1977) *The Politics of New Federalism*, Adelaide: Australian Political Studies Association, pp. 1–14.

Vishnyakov, V. (2002) 'Ukrupnenie ne Sama Tsel', *Rossiiskaya Federatsiya Segodnya*, No. 8, 2002.

Volkov, V. (2002) 'The Selective Use of State Capacity in Russia's Economy: Property Disputes and Enterprise Takeovers after 2000', Washington DC: PONARS Policy Memo No. 273, October. Online. Available . (accessed 20 April 2009).

Vyzhutovich, V. (2007) 'Dolgoe Proshchanie', Publication of the Moscow Centre for Political Technology, 7 August. Online. Available at politcom.ru/article.php?id=4926 (accessed 20 December 2007).

Waterman, H. (1969) *Political Change in Contemporary France. The Politics of an Industrial Democracy*, Columbus, Ohio: Charles E. Merrill.

Watts, R. (1966) *New Federations: Experiments in the Commonwealth*, Oxford: Clarendon Press.

—— (1989) 'Executive Federalism: The Comparative Perspective', in Shugarman, D. and Whitaker, R. (eds) (1989) *Federalism and Political Community*, Ontario: Broadview Press, pp. 439–61.

—— (1999) *Comparing Federal Systems*, 2nd edn, Montreal and Kingston: McGill-Queen's University Press.

Welch, S. (1993) *The Concept of Political Culture*, New York: St. Martin Press.

Wheare, K.C. (1963) *Federal Government*, 4th edn, London: Oxford University Press.

White, S. (1977) 'The USSR: Patterns of Autocracy and Industrialism', in Brown, A. and Gray, J. (eds) (1977) *Political Culture and Political Change in Communist States*, London: MacMillan, pp. 25–66.

Whitefield, S. (2005) 'Culture, Experience, and State Identity: A Survey-Based Analysis of Russians, 1995–2003', in Whitefield, S. (ed.) (2005) *Political Culture and Post-Communism*, London: Palgrave Macmillan, pp. 125–48.

Yorke, A. (2003) 'Business and Politics in Krasnoyarsk Krai', *Europe-Asia Studies*, 55(1): 241–62.

Zakharov, A. (ed.) (1997) *Asymmetrichnost Federatsii*, Moscow: TACIS.

Zubarevich, N. (2002) 'Prishel, Uvidel, Pobedil?', *Pro et contra*, 7(1): 107–19.

—— (2005) *Krupnyi Biznes v Regionakh Rossii: Territorialnye Strategii Razvitiya i Sotsialnye Interesy*, Moscow: Pomatur.

—— (2007) 'Sshit prostranstvo', *Ekspert*, 27 December.

Zubarevich, N., Petrov, N., and Titkov, A. (2001) 'Federalnye Okruga: 2000', in Petrov, N. (ed.) (2001) *Regiony Rossii v 1999: Ezhegodnoe Prilozhenie k 'Politicheskomu Almanakhu Rossii'*, Moscow: Gendalf, pp. 173–96.

Zudin, A. (2002) 'Rezhim Putina: Kontury Novoi Politicheskoi Sistemy', Moscow Centre for Political Technology, 17 November. Online. Available at www.cpt.ru/statint.php/ (accessed 20 April 2009).

—— (2005a) 'Vzaimootnosheniya krupnogo biznesa i vlasti pri Vladimire Putine i ikh vliyanie na situatsiyu v regionakh', in Frukhtman, J. (ed.) (2005) *Regionalnaya Elita v Sovremennoi Rossii*, Moscow: Fond Liberalnaya Missiya Press, pp. 37–64.

—— (2005b) *Biznes, Assotsiatsii i Gosudarstvo: Sravnitelnyi Analiz Razvitiya Otnoshenii na Zapade i v 'Postsotsialisticheskikh Stranakh'*, Moscow Centre for Political Technology, unpublished.

Zvonarev, Yu. (2007) 'V rezhime razminki', *Ekspert*, 23 April.

Index